Student Solutions Manual

Applied Statistics for Engineers and Scientists

THIRD EDITION

Jay Devore
California Polytechnic State University,
San Luis Obispo

Nicholas Farnum
California State University, Fullerton

Jimmy Doi
California Polytechnic State University,
San Luis Obispo

Prepared by

Soma Roy
California Polytechnic State University,
San Luis Obispo

Australia • Brazil • Mexico • Singapore • United Kingdom • United States

ISBN-13: 978-1-133-49218-4
ISBN-10: 1-133-49218-5

Cengage Learning
200 First Stamford Place, 4th Floor
Stamford, CT 06902
USA

Cengage Learning is a leading provider of customized learning solutions with office locations around the globe, including Singapore, the United Kingdom, Australia, Mexico, Brazil, and Japan. Locate your local office at: **www.cengage.com/global**.

Cengage Learning products are represented in Canada by Nelson Education, Ltd.

To learn more about Cengage Learning Solutions, visit **www.cengage.com**.

Purchase any of our products at your local college store or at our preferred online store **www.cengagebrain.com**.

Printed in the United States of America
1 2 3 4 5 6 7 17 16 15 14 13

TABLE OF CONTENTS

Chapter 1
Data and Distributions

Section 1.2

1. (a) MINITAB generates the following stem-and-leaf display of this data:

```
Stem-and-leaf of C1          N  = 27
Leaf Unit = 0.10

      1     5 9
      6     6 33588
   (11)     7 00234677889
     10     8 127
      7     9 077
      4    10 7
      3    11 368
```

The left most column in the MINITAB printout shows the cumulative numbers of observations from each stem *to the nearest tail* of the data. For example, the 6 in the second row indicates that there are a total of 6 data points contained in stems 6 and 5. MINITAB uses parentheses around 11 in row three to indicate that the **median** (described in Chapter 2, Section 2.1) of the data is contained in this stem. A value close to 8 is representative of this data.

What constitutes large or small variation usually depends on the application at hand, but an often-used rule of thumb is: the variation tends to be large whenever the spread of the data (the difference between the largest and smallest observations) is large compared to a representative value. Here, 'large' means that the percentage is closer to 100% than it is to 0%. For this data, the spread is 11 - 5 = 6, which constitutes 6/8 = .75, or, 75%, of the typical data value of 8. Most researchers would call this a large amount of variation.

(b) The data display is not perfectly symmetric around some middle/representative value. There tends to be some positive skewness in this data.

(c) In Chapter 1, outliers are data points that appear to be *very* different from the pack. Looking at the stem-and-leaf display in part (a), there appear to be no outliers in this data. (Chapter 2 gives a more precise definition of what constitutes an outlier).

(d) From the stem-and-leaf display in part (a), there are 3 leaves associated with the stem of 11, which represent the 3 data values that greater than or equal to 11. 10.7, which is represented by the stem of 10 and the leaf of 7, also exceeds 10. Therefore, the proportion of data values that exceed 10 is 4/27 = .148, or, about 15%.

3. A MINITAB stem-and-leaf display of this data is:

```
Stem-and-leaf of C1          N  = 36
Leaf Unit = .01

   1    3 1
   6    3 56678
  18    4 000112222234
  18    4 5667888
  11    5 144
   8    5 58
   6    6 2
   5    6 6678
   1    7
   1    7 5
```

Another method of denoting the pairs of stems having equal values is to denote the first stem by L, for 'low', and the second stem by H, for 'high'. Using this notation, the stem-and-leaf display would appear as follows:

```
3L   1
3H   56678
4L   000112222234
4H   5667888
5L   144
5H   58
6L   2
6H   6678
7L
7H   5
```

The stem-and-leaf display shows that .45 is a good representative value for the data. In addition, the display is not symmetric and appears to be positively skewed. The spread of the data is .75 - .31 = .44, which is.44/.45 = .978, or about 98% of the typical value of .45. Using the same rule of thumb as in Exercise 1, this constitutes a reasonably large amount of variation in the data. The data value .75 is a possible outlier (the definition of 'outlier' in Section 2.3, shows that .75 could be considered to be a 'mild' outlier).

Because the stem-and-leaf display is nearly symmetric around 90, a representative value of about 90 is easy to discern from the diagram. The most apparent features of the display are its approximate symmetry and the tendency for the data values to stack up around the representative value in a bell-shaped curve. Also, the spread of the data, 100.3-83.4 = 16.9 is a relatively small percentage (16.9/90 ≈ .18, or 18%) of the typical value of 90.

5. (a) Two-digit stems would be best. One-digit stems would create a display with only 2 stems, 6 and 7, which would give a display without much detail. Three-digit stems would cause the display to be much too wide with many gaps (stems with no leaves).

(b) The stem-and-leaf display below does not give up (truncate) the rightmost digit in the data:

```
64   33 35 64 70
65   06 26 27 83
66   05 14 94
67   00 13 45 70 70 90 98
68   50 70 73 90
69   00 04 27 36
70   05 11 22 40 50 51
71   05 13 31 65 68 69
72   09 80
```

(c) A MINITAB stem-and-leaf display of this data appears below. Note that MINITAB does truncate the rightmost digit in the data values.

```
  4    64 3367
  8    65 0228
 11    66 019
 18    67 0147799
 (4)   68 5779
 18    69 0023
 14    70 012455
  8    71 013666
  2    72 08
```

This display tends to be about as informative as the one in part (b). With larger sample sizes, the work involved in creating the display in part (c) would be much less than that required in part (b). In addition, for a larger sample size, the 'full' display in (b) would require a lot of room horizontally on the page to accommodate all the 2-digit leaves.

7. (a)

Number Nonconforming	Frequency	RelativeFrequency(Freq/60)
0	7	0.117
1	12	0.200
2	13	0.217
3	14	0.233
4	6	0.100
5	3	0.050
6	3	0.050
7	1	0.017
8	1	0.017
doesn't add exactly to 1 because relative frequencies have been rounded →		1.001

(b) The number of batches with at most 5 nonconforming items is $7+12+13+14+6+3=55$, which is a proportion of 55/60 = .917. The proportion of batches with (strictly) fewer than 5 nonconforming items is 52/60 = .867. Notice that these proportions could also have been computed by using the relative frequencies: e.g., proportion of batches with 5 or fewer nonconforming items = $1-(.05+.017+.107)=.916$; proportion of batches with fewer than 5 nonconforming items = $1-(.05+.05+.017+.107)=.866$.

(c) The following is a MINITAB histogram of this data. The center of the histogram is somewhere around 2 or 3, and it shows that there is some positive skewness in the data. Using the rule of thumb in Exercise 1, the histogram also shows that there is a lot of spread/variation in this data.

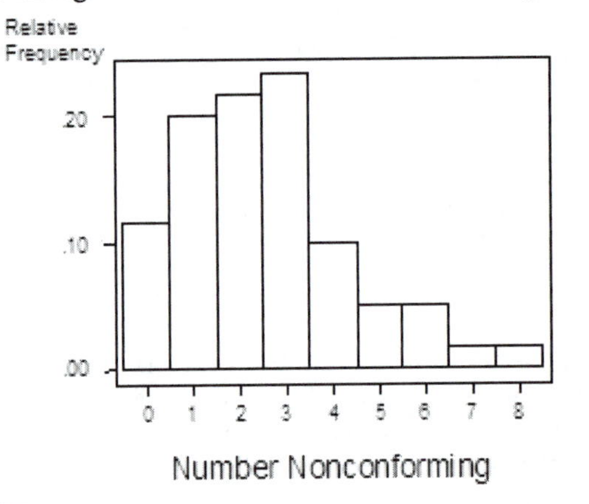

9. (a) From this frequency distribution, the proportion of wafers that contained at least one particle is (100-1)/100 = .99, or 99%. Note that it is much easier to subtract 1 (which is the number of wafers that contain 0 particles) from 100 than it would be to add all the frequencies for 1, 2, 3,... particles. In a similar fashion, the proportion containing at least 5 particles is (100 - 1-2-3-12-11)/100 = 71/100 = .71, or, 71%.

(b) The proportion containing between 5 and 10 particles is (15+18+10+12+4+5)/100 = 64/100 = .64, or 64%. The proportion that contain strictly between 5 and 10 (meaning strictly *more* than 5 and strictly *less* than 10) is (18+10+12+4)/100 = 44/100 = .44, or 44%.

(c) The following histogram was constructed using MINITAB. The data was entered using the same technique mentioned in the answer to exercise 8(a). The histogram is *almost* symmetric and unimodal; however, it has a few relative maxima (i.e., modes) and has a very slight positive skew.

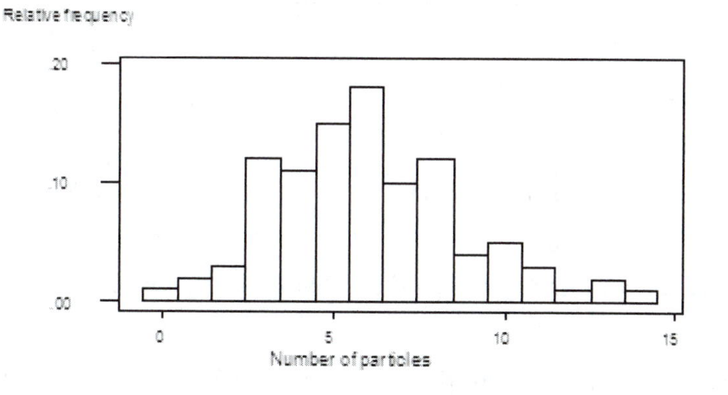

11. (a) The following stem-and-leaf display was constructed using MINITAB:

```
Stem-and-leaf of C1          N  = 47
Leaf Unit = 100

  12     0 123334555599
  23     1 00122234688
 (10)    2 1112344477
  14     3 0113338
   7     4 37
   5     5 23778
```

A typical data value is somewhere in the low 2000's. The display almost unimodal (the stem at 5 would be considered a mode, the stem at 0 another) and has a positive skew.

(b) A histogram of this data, using classes of width 1000 separated at 0, 1000, 2000, and 6000 is shown below. The proportion of subdivisions with total length less than 2000 is (12+11)/47 = .489, or 48.9%.

Between 2000 and 4000, the proportion is (10 + 7)/47 = .362, or 36.21%. The histogram shows the same general shape as depicted by the stem-and-leaf display in part (a).

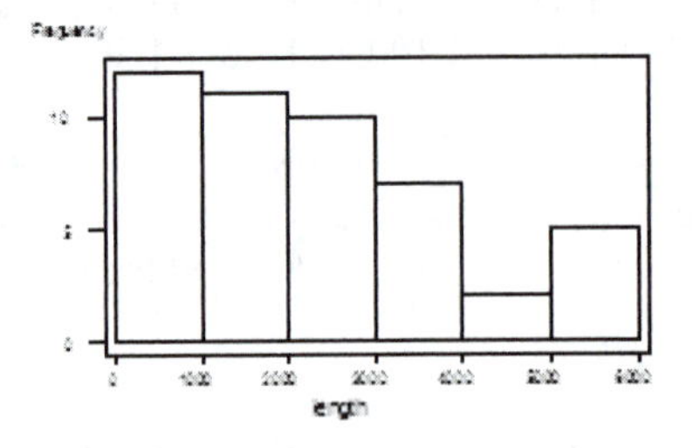

13. (a) Proportion of herds with only one giraffe = 589/1570 = 0.3752

(b) Proportion of herds with six or more giraffes = (89+57+…+ 1 + 1)/1570 or 1 – (589 + 190 + 176 + 157 + 115)/1570 = 0.2185

(c) Proportion of herds that had between 5 and 10 giraffes, inclusive = (115+89+57+55+33+31)/1570 = 0.242

(d) The distribution of herd size is skewed to the right, with very few large herds, and majority of herds being smaller than 3 to 4 in size.

15.

(a) Yes: the proportion of sampled angles smaller than 15° is .177 + .166 + .175 = .518.

(b) The proportion of sampled angles at least 30° is .078 + .044 + .030 = .152.

(c) The proportion of angles between 10° and 25° is roughly .175 + .136 + (.194)/2 = .408.

(d) The distribution of misorientation angles is heavily positively skewed. Though angles can range from 0° to 90°, nearly 85% of all angles are less than 30°. Without more precise information, we cannot tell if the data contain outliers.

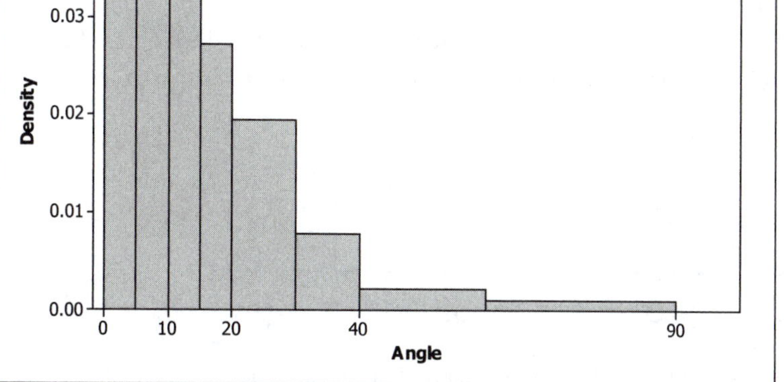

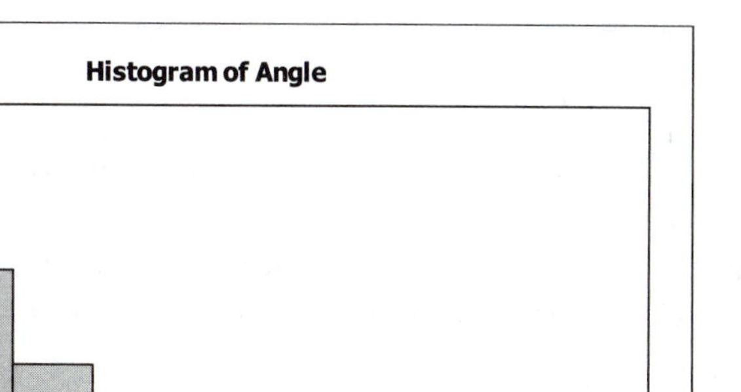

17. The histogram of this data appears below. A typical value of the shear strength is around 5000 lb. The histogram is almost symmetric and approximately bell-shaped.

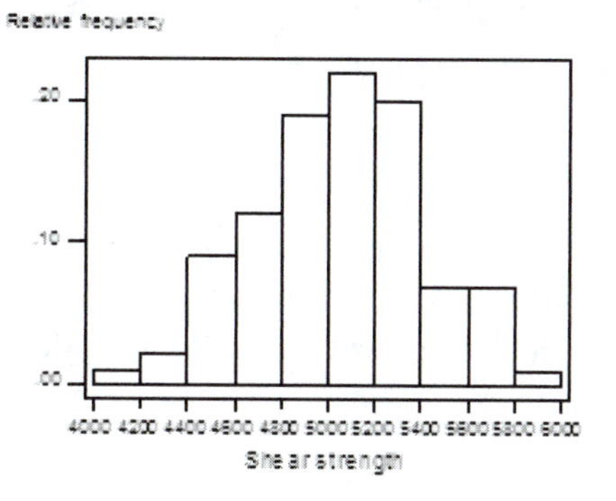

Section 1.3

19. (a) The density curve forms a rectangle over the interval [4, 6]. For this reason, uniform densities are also called **rectangular densities** by some authors. Areas under uniform densities are easy to find (i.e., no calculus is needed) since they are just areas of rectangles. For example, the total area under this density curve is $\frac{1}{2}(6-4) = 1$.

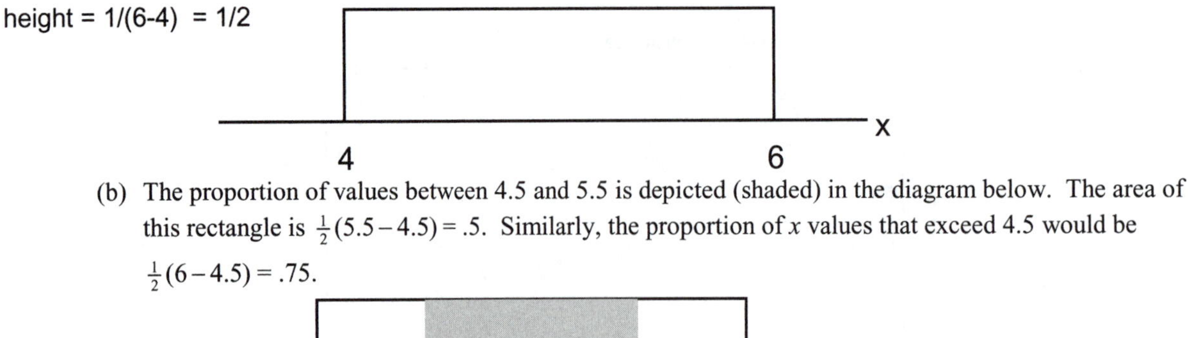

(b) The proportion of values between 4.5 and 5.5 is depicted (shaded) in the diagram below. The area of this rectangle is $\frac{1}{2}(5.5-4.5)=.5$. Similarly, the proportion of x values that exceed 4.5 would be $\frac{1}{2}(6-4.5)=.75$.

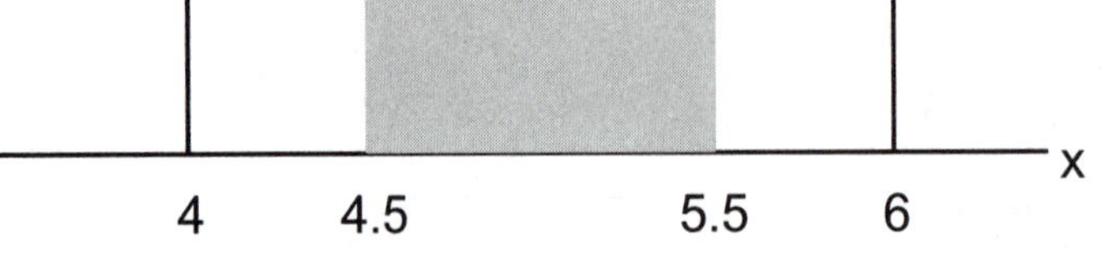

(c) The median of this distribution is 5 because exactly half the area under this density sits over the interval [4,5].

(d) Since 'good' processing times are short ones, we need to find the particular value x_0 for which the proportion of the data less than x_0 equals .10. That is, the area under the density to the left of x_0 must equal .10. Therefore, the area $=.10=\frac{1}{2}(x_0-4)$, and so $x_0-4=.20$. Thus, $x_0=4.20$.

21. (a) The density curve forms an isosceles triangle over the interval [0, 10]. For this reason, such densities are often called **triangular densities**. The total area under this density curve is simply the area of the triangle, which is $\frac{1}{2}$ (base)(height) = $\frac{1}{2}$ (10)(.2) = 1. The height of the triangle is the value of f(x) at x = 5; i.e., f(5) = .4 -.04(5) = .2.

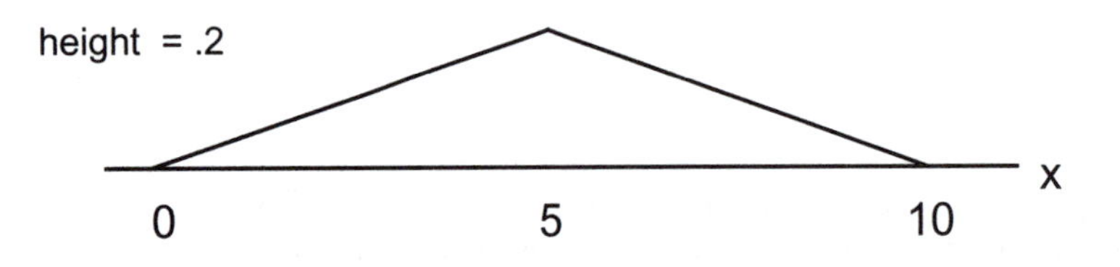

(b) Proportion $(x \le 3)$ = $\frac{1}{2}(3-0)f(3)$ = $\frac{1}{2}$ (3)(.12) = .18.

Proportion $(x \ge 7)$ = $\frac{1}{2}(10-7)f(7)$ = $\frac{1}{2}$ (3)(.12) = .18.

Proportion $(x \ge 4)$ = 1 – Proportion $(x < 4)$ = $1-\frac{1}{2}(4-0)f(4)$ = 1 - $\frac{1}{2}$ (4)(.16) = .68.

Proportion $(4 < x < 7)$ = 1 - [Proportion $(x \le 4)$ + Proportion $(x \ge 7)$] = 1 - [.32+.18] = .50.

(c) This particular value x_0 will have 10% of the waiting times above it, so Proportion $(x < x_0) = .10$. That is, $\frac{1}{2}(10 - x_0)f(x_0) = .10$. Since $f(x) = .4 - .04x$ in this region, we have $.10 = \frac{1}{2}(10 - x_0)(.4 - .04x_0)$. Multiplying the factors and collecting terms yield a quadratic equation $.04x_0^2 - .8x_0 + 3.8 = 0$, whose solutions are 7.7639 and 12.236. Of these values, only $x_0 = 7.7639$ lies within the interval [0,10].

Alternate solution: Notice that f(x) is symmetric around x = 5, so that the triangle that captures the leftmost 10% of the area will have a base x_1 of the *same* length as the triangle that captures the rightmost 10%. The left-hand triangular region is much easier to work with: area $= .10 = \frac{1}{2}(x_1)f(x_1) = \frac{1}{2}(x_1)(.04x_1)$, so $x_1^2 = 5$ and $x_1 = +\sqrt{5}$. Therefore, $x_0 = 10 - x_1 = 7.7639$.

23. (a)

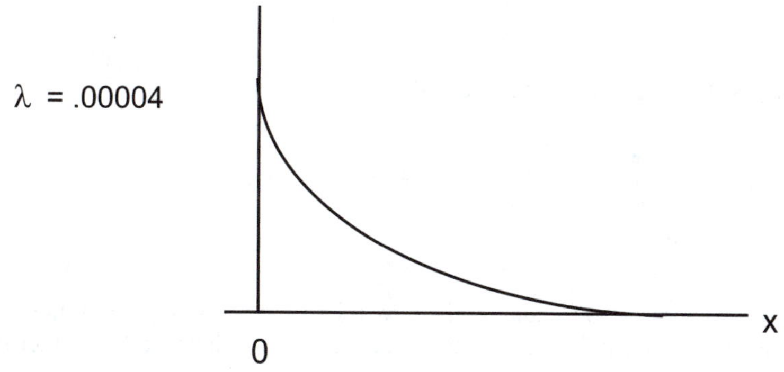

λ = .00004

0

x

(b) $$\int_{20,0000}^{\infty} .00004e^{-.00004x}\,dx = \left[\frac{-(.00004)e^{-.00004x}}{(.00004)}\right]_{20,000}^{\infty} = -e^{-.00004x}\Big|_{20,000}^{\infty}$$

$$= 0 - (-e^{-.00004(20,000)}) = e^{-.8} = .449.$$

Note: for any exponential density curve, the area to the right of some fixed constant always equals $e^{-\lambda c}$, as our integration above shows. That is,

$$\text{Proportion } (x > c) = \int_{c}^{\infty} \lambda e^{-\lambda x}\,dx = e^{-\lambda c}.$$

We will use this fact in the remainder of the chapter instead of repeating the same type of integration as in part (a)

Proportion ($x \le 30{,}000$) = 1 - Proportion ($x > 30{,}000$) $= 1 - e^{-\lambda c} =$
$1 - e^{-.00004(30{,}000)} = 1 - e^{-1.2} = .699$.

Proportion ($20{,}000 \le x \le 30{,}000$) = Proportion ($x > 30{,}000$) - Proportion($x \le 20{,}000$) = .699 - (1-.449) = .148.

(c) For the best 1%, the lifetimes must be at least x_0, where Proportion($x \ge x_0$) = .01, which becomes $e^{-\lambda x_0} = .01$. Taking natural logarithms of both sides, $-\lambda x_0 = \ln(.01)$, so $x_0 = -\ln(.01)/\lambda$ = 4.60517/.00004 = 115,129.25. For the worst 1%, we have Proportion($x \le x_0$) = .01, which is equivalent to saying that Proportion($x \ge x_0$) = .99, so $e^{-\lambda x_0} = .99$. Taking logarithms, $-\lambda x_0 = \ln(.99)$, so $x_0 = -\ln(.99)/\lambda = 251.26$.

25. (a) The total area under the density curve must equal 1, so:

$$1 = \int_2^4 c[1-(x-3)^2]\,dx = c\int_2^4 [1-(x-3)^2]\,dx = c\left[\left(x - \tfrac{1}{3}(x-3)^3\right)\right]_2^4$$

$$= c[(4-\tfrac{1}{3}) - (2+\tfrac{1}{3})] = \tfrac{4}{3}c, \text{ so, } c = \tfrac{3}{4}.$$

(b) Proportion($x > 3$) equals .50. No calculation is needed if you draw a diagram of the density curve and notice that it is symmetric around $x = 3$, which means that exactly half the area under the curve lies on either side of $x = 3$.

(c) Proportion($3-.25 \le x \le 3+.25$) = Proportion($2.75 \le x \le 3.25$) =

$$\int_{2.75}^{3.25} f(x)dx = \frac{3}{4}\left[x - \frac{(x-3)^3}{3}\right]_{2.75}^{3.25} = .367.$$

27. (a) Proportion($x \le 3$) = .10 + .15 + .20 + .25 = .70.
Proportion($x < 3$) = Proportion($x \le 2$) = .10 + .15 + .20 = .45.

(b) Proportion($x \ge 5$) = 1 - Proportion($x < 5$) = 1 - (.10+.15+.20+.25+.20) = .10.

(c) Proportion($2 \le x \le 4$) = .20 + .25 + .20 = .65

(d) At least 4 lines will *not* be in use whenever 3 or fewer lines *are* in use. At most 3 lines are in use .70, or 70%, of the time from part (a) of this exercise.

29. The 10 possible samples of size two are shown below along with the corresponding values of x:

sample →	{1,2}	{1,3}	{1,4}	{1,5}	{2,3}	{2,4}	{2,5}	{3,4}	{3,5}	{4,5}
x →	2	1	1	1	1	1	1	0	0	0

The mass function for x is then:

x →	0	1	2
p(x) →	3/10	6/10	1/10

Section 1.4

31. (a) Proportion($z \le 1.78$) = .9625 (Table I)

(b) Proportion($z > .55$) = 1 - Proportion($z \le .55$) = 1 - .7088 = .2912.

(b) Proportion($z > -.80$) = 1 - Proportion($z \le -.80$) = 1 - .2119 = .7881.

(c) Proportion($.21 \le z \le 1.21$) = Proportion($z \le 1.21$) - Proportion($z \le .21$) = .8869 - .5832 = .3037.

(d) Proportion($z \le -2.00$ or $z \ge 2.00$) = Proportion ($z \le -2.00$) + [1- Proportion($z < 2.00$)] = .0228 + [1 - .9772] = .0456. Alternatively, using the fact that the z density is symmetric around z = 0, Proportion($z \le -2.00$) = Proportion($z \ge 2.00$), so the answer is simply 2 Proportion($z \le -2.00$) = 2(.0228) = .0456.

(e) Proportion($z \le -4.2$) = .0000

(f) Proportion($z > 4.33$) = .0000

33. (a) Let z^* denote the value of z that is exceeded by the largest 15% of all z values. Then , Proportion($z > z^*$) = .15. In terms of left-tail areas, Proportion($z \le z^*$) = .85. From Table I, Proportion($z \le 1.04$) = .8508 and Proportion($z \le 1.03$) = .8485, so, using linear interpolation, $z^* = 1.03 + [(.8500-.8485)/(.8508-.8485)](1.04-1.03) = 1.036$. That is, any z value greater than 1.036 (or approximately 1.04) will be in the top 15%.

(b) Proportion($z \leq z^*$) = .25. From Table I, Proportion($z \leq -.68$) = .2483 and Proportion($z \leq -.67$)= .2514, so interpolation gives $z^* = -.67 + [(.2500-.2483)/(.2514-.2483)](-.68-(-.67)) = -.6755$ or, approximately, $z^* = -.675$.

(c) The 4% farthest from 0 consists of the 2% in the upper tail and the 2% in the lower tail. In the upper tail, Proportion($z > z^*$) = .02, which can be re-expressed as a left-tail area Proportion($z \leq z^*$) = .98. From Table I, Proportion($z \leq 2.05$) = .9798 and Proportion($z \leq 2.06$) = .9803, so $z^* \approx 2.055$. Similarly, a z^* value of -2.055 will capture the lower 2% of the z values.

35. (a) Let x = bolt thickness. Then, the statement $x \leq 11$ describes bolts that are less than one σ (since $\sigma = 1$) to the right of $\mu = 10$. The equivalent area under the z density is $z \leq 1$ (since $\mu = 0$ and $\sigma = 1$). From Table I, this area/proportion is .8413.

(b) The region $7.5 \leq x \leq 12.5$ describes bolts whose lengths are within 2.5 σ's from the mean of 10. The equivalent z curve area is described by $-2.5 \leq z \leq 2.5$. From Table I, this area equals Proportion($z \leq 2.5$) - Proportion($z \leq -2.5$) = .9938 - .0062 = .9876.

(c) The region $x > 11.5$ describes bolts that are more than 1.5 σ's above the mean. The corresponding z curve region is $z > 1.5$. From Table I, Proportion($z > 1.5$) = 1 - Proportion($z \leq 1.5$) = 1 - .9332 = .0668.

37. (a) Let x = maximum moped speed. Then, $x \leq 50$, when standardized, becomes $z \leq (50-46.8)/1.75 = 1.83$. From Table I, Proportion($z \leq 1.83$) = 0.9664.

(b) Similarly, $x \geq 48$ standardizing yields $z \geq (48-46.8)/1.75 = 0.69$. From Table I, Proportion($z > 0.69$) = 1 - Proportion($z \leq 0.69$) = 1- .755 = .245.

(c) We need to find the value of x^* for which Proportion($x > x^*$) = .75, or equivalently, Proportion($x \leq x^*$) = .25. The corresponding statement about a z curve is Proportion($z \leq z^*$) = .25. From Table I, $z^* \approx -.675$; i.e., z^* is .675 σ's below the mean. So, x^* must be .675 σ's below the mean of the x data: $x^* = 46.8 - .675(1.75) = 45.62$km/h.

39. (a) Let x = substrate concentration. Standardizing $x > .25$ yields $z > (.25-.30)/.06$, or $z > -.83$. From Table I, Proportion($z > -.83$) = 1 - Proportion($z \leq -.83$) = 1 - .2033 = .7967.

(b) $x \leq .10$ standardizes to $z \leq (.10-.30)/.06 = -3.33$, so Proportion($z \leq -3.33$) = .0004.

(c) Let x^* be the value exceeded by the largest 5% of x values. Then Proportion($x > x^*$) = .05. The equivalent statement about the z density is Proportion($z > z^*$) = .05, or, in terms of left-tail areas, Proportion($z \leq z^*$) = .95. From Table I, $z^* = 1.645$, which is 1.645 σ's above the mean of the z data. Therefore, x^* must be 1.645 σ's above the mean of the x data: $x^* = .30 + 1.645(.06) = .399$.

41. (a) Because x is a discrete variable, the best approximation to the Proportion($20 \le x \le 40$) is to find the area under the x density curve between the points 19.5 and 40.5. That is, Proportion($(19.5-25)/5 \le z \le (40.5-25)/5$) = Proportion($-1.1 < z < 3.1$) = Proportion($z < 3.1$) - Proportion($z < -1.1$) = .9990 - .1357 = .8633.

(b) Using the same sort of approximation as in part (a), Proportion($x \le 30.5$) = Proportion($z \le (30.5-25)/5$) = Proportion($z \le 1.1$) = .8643. Similarly, the proportion of reels having *fewer* than 30 flaws is approximated by Proportion($x < 29.5$) = Proportion($z \le (29.5-25)/5$) = Proportion($z \le .9$) = .8159.

Section 1.5

43. (a) Let x = ductile strength. Then, Proportion($x > 120$) = Proportion($\ln(x) > \ln(120)$) = Proportion($\ln(x) > 4.78749$). Because ln(x) has a normal distribution with $\mu = 5$ and $\sigma = .1$, we can then standardize to find Proportion($\ln(x) > 4.78749$) = Proportion($z > (4.78749-5)/.1$) = Proportion($z > -2.13$) = 1 - Proportion($z < -2.13$) = 1 - .0166 = .9834. Because x has a continuous distribution, Proportion($x \ge 120$) also equals .9834.

(b) Proportion($110 \le x \le 130$) = Proportion($\ln(110) \le \ln(x) \le \ln(130)$) = Proportion($4.70048 \le \ln(x) \le 4.86753$) = Proportion($(4.70048-5)/.1 \le z \le (4.86753-5)/.1$) = Proportion($-3.00 \le z \le -1.33$) = Proportion($z < -1.33$) - Proportion($z < -3.00$) = .0918 - .0013 = .0905.

(c) Let x^* be the minimum acceptable strength. Then, Proportion($x \le x^*$) = .05. Taking logarithms and then standardizing, .05 = Proportion($\ln(x) \le \ln(x^*)$) = Proportion($z \le (\ln(x^*)-5)/.1$). From Table I, Proportion($z \le -1.645$) $\approx$ .05, so $(\ln(x^*)-5)/.1 = -1.645$ and therefore $\ln(x^*) = 5-1.645(.1) = 4.8355$. Exponentiating both sides gives $x^* = \exp(4.8355) = 125.90$.

45. (a) Let x = tensile strength. A graph of the density function appears below:

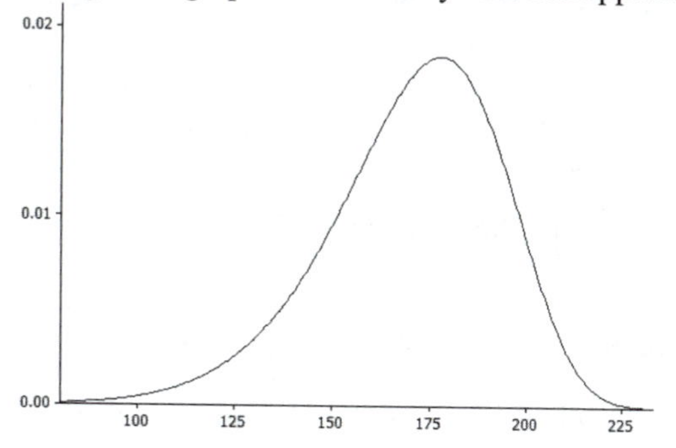

Note: The mean of a Weibull distribution is $\mu = \beta\Gamma(1+1/\alpha)$, where $\Gamma(\cdot)$ denotes the **gamma function** (whose values can be found by using Excel or most statistical packages. For example, in this exercise, $\mu = 180\Gamma(1+1/9) = 180\Gamma(1.1111) = 180(.946968) = 170.45$.

(b) Proportion(x > 175) = $e^{-(175/180)^9}$ = .4602.

(c) Proportion(150 ≤ x ≤ 175) = $\text{Proportion}(x \le 175) - \text{Proportion}(x \le 150) =$

$(1-e^{-(175/180)^2}) - (1-e^{-(150/180)^2} = .3636$

(d) Let x* denote the value that exceeds the weakest 10% of the data. Then, .10 = Proportion(x ≤ x*) = 1- $e^{-(x^*/180)^9}$, or $e^{-(x^*/180)^9}$ = .90. Taking logarithms of both sides, -.1053605 = ln(.90) = $-(x^*/180)^9$, so .1053605 = $(x^*/180)^9$, which means that x*/180 = $.1053605^{1/9}$ = .778769 and, finally, x* = 180(.778769) = 140.18.

47. (a) Let x = tree diameter. Then, Proportion(2 ≤ x ≤ 4) = Proportion(x > 2) - Proportion(x > 4)= $e^{-((2-1.3)/5.8)^4}$ - $e^{-((4-1.3)/5.8)^4}$ = .999788 - .954124 = .0456.

(b) Proportion(x ≥ 5) = $e^{-((5-1.3)/5.8)^4}$ = .8474.

(c) Let $\tilde{x}$ denote the median tree diameter. Then, .50 = Proportion(x ≥ $\tilde{x}$) = $e^{-((\tilde{x}-1.3)/5.8)^4}$. Taking logarithms of both sides, -.6931472 = ln(.50) = $-((\tilde{x}-1.3)/5.8)^4$, so, $\tilde{x}$ = 1.3 + 5.8$(.6931472)^{1/4}$ = 6.592.

49. (a) Let x denote the drill lifetimes. Because x is lognormal, $\ln x$ follows a normal distribution with parameters $\mu = 4.5$ and $\sigma = .8$. Therefore:

$$\text{Proportion}(x \le 100) = \text{Proportion}(\ln x \le \ln 100) = \text{Proportion}\left(\frac{\ln x - \mu}{\sigma} \le \frac{\ln 100 - \mu}{\sigma}\right)$$

$$= \text{Proportion}\left(z \le \frac{\ln 100 - 4.5}{.8}\right) = \text{Proportion}(z \le .131) \approx .5517.$$

(b) The proportion of lifetime values that are at least 200 is given by:

$$\text{Proportion}(x \ge 200) = \text{Proportion}(\ln x \ge \ln 200) = \text{Proportion}\left(\frac{\ln x - \mu}{\sigma} \ge \frac{\ln 200 - \mu}{\sigma}\right)$$

$$= \text{Proportion}\left(z \geq \frac{\ln 100 - 4.5}{.8}\right) = \text{Proportion}(z \geq 1.00) = 1 - \text{Proportion}(z < 1.00) = 1 - .8413 = .1587.$$

Because this is the continuous distribution, the proportion of values that are at least 200 is the same as the proportion of values that exceed 200: that is, .1587, or 15.87%.

51. (a) Let x = 1-hour significant wave height follow a Weibull distribution with parameters $\alpha = 1.817$ and $\beta = 0.863$.

$$P(\leq 0.5) = \int_0^{0.5} \frac{1.817}{0.863^{1.817}} x^{1.817-1} e^{-(x/0.863)^{1.817}} dx = \left[1 - e^{-(x/0.863)^{1.817}}\right]_0^{0.5} = 1 - e^{-(0.5/0.863)^{1.817}} = 1 - 0.6901 = 0.3099$$

(b)
$$\text{Proportion}(.2 \leq x \leq .6) = \left[1 - e^{-(x/.863)^{1.817}}\right]_{0.2}^{0.6}$$
$$= e^{-(.2/.863)^{1.817}} - e^{-(.6/.863)^{1.817}} \approx .3357$$

(c) Let $x*$ denote the 90th percentile of the lifetime distribution. Then,

$$.90 = \int_0^{x*} \frac{1.817}{0.863^{1.817}} x^{1.817-1} e^{-(x/0.863)^{1.817}} dx = \left[1 - e^{-(x/.863)^{1.817}}\right]_0^{x*} = 1 - e^{-(x*/.863)^{1.817}}.$$ Solve for

the value of $x*$ in the equation $.90 = 1 - e^{-(x*/.863)^{1.817}}$. This yields $e^{-(x*/.863)^{1.817}} = .10$, and so $\ln .10 = -(x*/.863)^{1.817}$. This implies that $x* = .863(-\ln .10)^{1/1.817} \approx 1.366$.

Similarly, let $x**$ denote the 10th percentile of this distribution. Then $.10 = 1 - e^{-(x**/.863)^{1.817}}$, and so $x** = .863(-\ln .90)^{1/1.817} \approx 0.2501$.

Section 1.6

53. (a) Let x = number of goblets with flaws in a box of 6. Then x has a binomial distribution with n = 6, π = .12. Using the formula for the binomial function, Proportion(x = 1) = $\frac{6!}{1!5!}(.12)^1(1-.12)^5$ = 6(.12)(.52773) = .3799, or, about .38.

(b) Proportion(x $\geq$ 2) = 1 - Proportion(x $\leq$ 1) = 1 - [Proportion(x =0) + Proportion(x =1)] = 1 - [$\frac{6!}{0!6!}(.12)^0(1-.12)^6 + \frac{6!}{1!5!}(.12)^1(1-.12)^5$] = 1 - [.4644 +.3799] = .1557.

(c) Proportion$(1 \le x \le 3)$ = Proportion$(x = 1)$ + Proportion$(x = 2)$ + Proportion$(x = 3)$ = $\frac{6!}{1!5!}(.12)^1(1-.12)^5$ + $\frac{6!}{2!4!}(.12)^2(1-.12)^4$ + $\frac{6!}{3!3!}(.12)^3(1-.12)^3$ = .3799 + .1295 + .0236 = .533.

55. (a) Let x = number of bits erroneously transmitted. Then, x is binomial with n = 20, π = .10, so Proportion$(x \le 2)$ = .122 + .270 + .285 = .677 (from Table II).

(b) Proportion$(x \ge 5)$ = .032 + .009 + .002 + .000 + ….+ .000 = .043.

(c) 'More than half' means 11 or more, so Proportion$(x \ge 11)$ = .000 + … + .000 = .000.

57. (a) Let x denote the number of drivers. Then, Proportion$(x \le 10)$ = .000 + .000 + …+ .001 + .003 + .006 = .01 (Table III, $\lambda = 20$).

(b) Proportion$(x > 20)$ = .085 + .077 + ….+ .013 + .008 = .428. (This answer of .428 is due to the limitation of the table. The actual answer should be .4409.)

(c) Proportion$(10 \le x \le 20)$ = .006 + .011 + … + .089 + .089 = .556.
Proportion$(10 < x < 20)$ = .011 +… + .089 = .461.

59. Using Table III ($\lambda = 20$), Proportion$(x \ge 15)$ = 1 - Proportion$(x \le 14)$ = 1 - (.000 + .000+ … + .001+ .001 + .003 + .006 + .011 + .018 + .-27 + .039) = 1 - .106 = .894. Similarly, Proportion$(x \le 25)$ = 1 - Proportion$(x \ge 26)$ = 1 - (.034 + .025 + .018 + .013 + .008) = 1 - .098 = .902.

Supplementary Exercises

61. (a) The histogram appears below. A representative value for this data would be x = 90. The histogram is reasonably symmetric, unimodal, and somewhat bell-shaped. The variation in the data is not small since the spread of the data (99-81 = 18) constitutes about 20% of the typical value of 90 (cf. Answer to Exercise 1 for a discussion of variation).

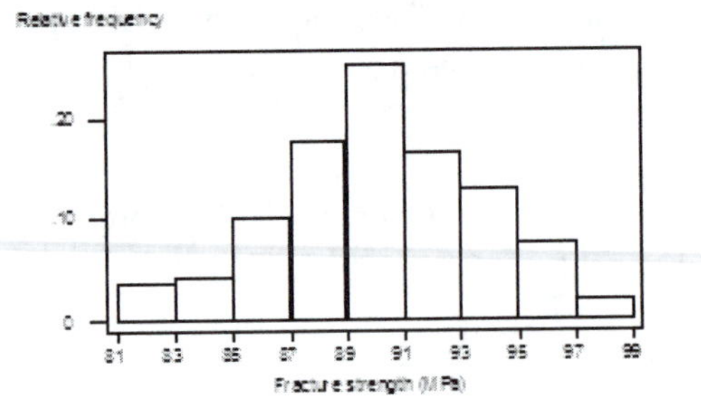

(b) The proportion of the observations that are at least 85 is 1 - (6+7)/169 = .9231. The proportion less than 95 is 1 - (22+13+3)/169 = .7751.

(c) x = 90 is the midpoint of the class 89-<91, which contains 43 observations (a relative frequency of 43/169 = .2544. Therefore, about half of this frequency, .1272, should be added to the relative frequencies for the classes to the left of x = 90. That is, the approximate proportion of observations that are less than 90 is .0355 + .0414 + .1006 + .1775 + .1272 = .4822.

63. (a) Let $y = x - 0.5$ follow a Weibull distribution with parameters $\alpha = 2.2$ and $\beta = 1.1$.
We want to find P(x > 1.5) = P(y + 0.5 > 1.5) = P(y > 1)

$$P(y>1)=\int_1^{\infty}\frac{2.2}{1.1^{2.2}}y^{2.2-1}e^{-(y/1.1)^{2.2}}dy=\left[1-e^{-(y/1.1)^{2.2}}\right]_1^{\infty}=e^{-(1/1.1)^{2.2}}=0.4445$$

(b) We want to find x* such that P(x > x*) = 0.10 ➔P(y > x* - 0.5) = 0.10

$$P(y>x^*-0.5)=\int_{x^*-.5}^{\infty}\frac{2.2}{1.1^{2.2}}y^{2.2-1}e^{-(y/1.1)^{2.2}}dy=\left[1-e^{-(y/1.1)^{2.2}}\right]_{x^*-.5}^{\infty}=e^{-((x^*-.5)/1.1)^{2.2}}=0.10$$

$$\Rightarrow -((x^*-.5)/1.1)^{2.2}=\ln(0.1)\Rightarrow x^*=(-\ln(0.1))^{1/2.2}*1.1+.5=2.107$$

Thus, the 90th percentile of the elapsed time distribution is 2.107days.

65. (a) Let x denote the unloading time. Since x is lognormal, then $\ln x$ is normal with mean 6.5 and standard deviation .75.

$$\text{Proportion }(x>1000)=\text{Proportion }(\ln x>\ln 1000)=\text{Proportion}\left(\frac{\ln x-6.5}{.75}>\frac{\ln 1000-6.5}{.75}\right)$$

$$=\text{Proportion }(z>.54)=1-\text{Proportion}(z<.54)=1-.7054=.2946$$

Proportion(x>2000) = Proportion (ln x > ln2000) = Proportion(z>1.47) = 0.0708;
Proportion(x>3000) = Proportion (ln x > ln3000) = Proportion(z>2.01) = 0.0222

(b) Proportion(2500< x<5000) = Proportion (ln(2500) < ln x < ln5000) = Proportion(1.77< z < 2.69) = 0.0348.

(c) Let x* be the value that characterizes the fastest 10% of all times. Proportion (x < x*) = 0.10, so Proportion(ln x < ln x*) = 0.10 ➔ Proportion (z < (ln x* - 6.5)/.75) = 0.10, so (ln x* - 6.5)/.75 = -1.282 ➔ ln x* = 6.5 – 1.282(.75) = 5.5385 ➔x* = exp(5.5385) = 254.3

(d) Yes, the positive (right) skewness is very pronounced.

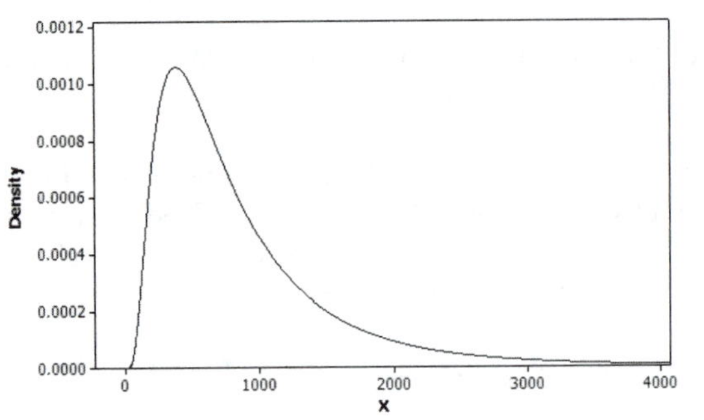

67. (a) Accommodating everyone who shows up means that $x \le 100$, so Proportion($x \le 100$) =.05 + .10 + … + .24 + .17 = .82.

(b) Proportion($x > 100$) = 1 - Proportion($x \le 100$) = 1 - .82 = .18.

(c) The first standby passenger will be able to fly as long as the number who show up is $x \le 99$, which leaves one free seat. This proportion is .65. The third person on the standby list will get a seat as long as $x \le 97$, where Proportion($x \le 97$) = .05 + .10 + .12 = .27.

69. (a) The area under the density must equal 1. To verify this, make the substitution $u = x^2/(2\theta^2)$, which gives $du = xdx/\theta^2$, so $\int_0^\infty \frac{x}{\theta^2} e^{-x^2/(2\theta^2)} dx = \int_0^\infty e^{-u} du = -e^{-u}\Big]_0^\infty = 0 - (-e^0) = 1$. For a finite upper limit x = c, the integral becomes: Proportion($x \le c$) = $\int_0^c \frac{x}{\theta^2} e^{-x^2/(2\theta^2)} dx = -e^{-u}\Big]_0^{c^2/(2\theta^2)} = 1 - e^{-c^2/(2\theta^2)}$.

(b) For $\theta = 100$, Proportion($x \le 200$) = $1 - e^{-200^2/(2(100)^2)} = 1 - e^{-2} = 1 - .135335 = .8647$.

Proportion($x \ge 200$) = 1 - Proportion($x < 200$) = 1 - .8647 = .1353.
Proportion($100 \le x \le 200$) = Proportion($x \le 200$) - Proportion($x \le 100$) = .8647 - $[1 - e^{-100^2/(2(100)^2)}]$ = .8647 - $[1 - e^{-.5}]$ = .4712.

71. (a) The density curve is shown below. In essence, the curve is just an exponential density that has been shifted to the right by 0.5 units.

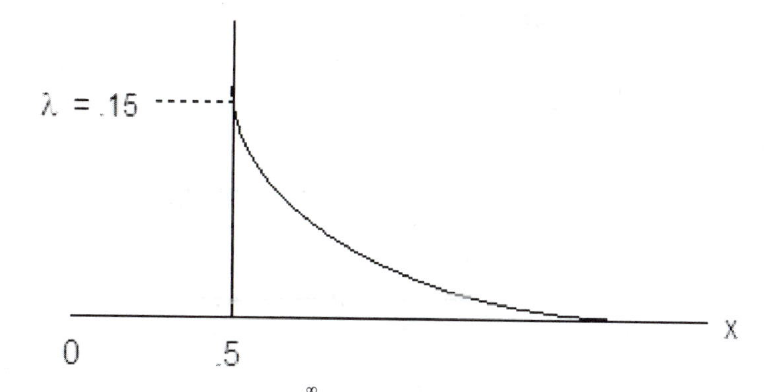

The total area under the density curve is $\int_{-\infty}^{\infty} f(x)dx = \int_{.5}^{\infty} .15e^{-.15(x-.5)}dx$. Using the substitution y = x-.5, the last integral becomes $\int_{0}^{\infty} .15e^{-.15y}dy$, which equals 1 because it is the area under an exponential density curve with λ = .15.

(b) Using the same substitution as in part (a), Proportion(x ≤ 5) = $\int_{.5}^{5} .15e^{-.15(x-.5)}dx$ = $\int_{0}^{4.5} .15e^{-.15y}dy$ = 1 - $e^{-.15(4.5)}$ = .491. Similarly, Proportion(5 < x < 10) = $\int_{5}^{10} .15e^{-.15(x-.5)}dx$ = $\int_{4.5}^{9.5} .15e^{-.15y}dy$ = $e^{-.15(4.5)}$ - $e^{-.15(9.5)}$ = .269.

(c) Let x^* denote the median of the headway times. Then, .50 = $\int_{.5}^{x^*} f(x)dx$ = $\int_{0}^{x^*-.5} .15e^{-.15y}dy = 1 - e^{-.15(x^*-.5)}$, so $e^{-.15(x^*-.5)}$ = 1 - .50 = .50. Taking logarithms of both sides we find -.15(x^*-.5) = ln(.50), so x^* = .5 + ln(.50)/(-.15) = 5.12.

(d) Let x^* denote the 90th percentile of the headway times. Then, .90 = $\int_{.5}^{x^*} f(x)dx$. Following the same steps as in part (c) we find x^* = .5 +ln(1-.90)/(-.15) = 15.85.

73. (a) The cumulative proportion graph is shown below:

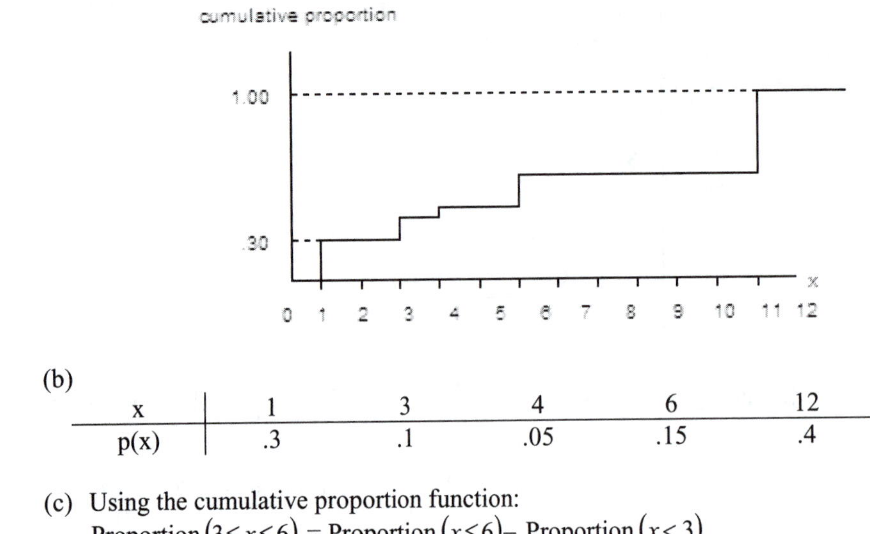

(b)

x	1	3	4	6	12
p(x)	.3	.1	.05	.15	.4

(c) Using the cumulative proportion function:
Proportion $(3 \le x \le 6)$ = Proportion $(x \le 6)$ – Proportion $(x < 3)$
= .60 - .30 = .30

Using the mass function:
Proportion $(3 \le x \le 6) = p(3) + p(4) + p(6) = .1 + .05 + .15 = .30$
We do obtain the same answers.

75. Let x = bursting strength is normally distributed with mean 250psi and standard deviation 30psi.
Proportion(x > 300) = Proportion(z > (300-250)/30) = Proportion(z > 1.67) = 0.0475

Let Y = number of bottles in carton of 12 that'll have bursting strength exceed 300; Y is binomial with n = 12 and $\pi = 0.0475$.
Then Proportion $(Y \ge 1) = 1 -$ Proportion$(Y<1) = 1 -$ Proportion$(Y=0) = 1 - (0.0475)^0\,(1-0.0475)^{12} = 1 - 0.5577 = 0.4423$.

Chapter 2
Numerical Summary Measures

Section 2.1

1. (a) The sample mean, $\bar{x}$ = 640.5 (that is, $640500), and the sample median, $\tilde{x}$ = 582.5 (that is, $582500). The average price of houses sold in the previous month is $640,500, whereas about 50% of the houses sold in the previous month sold for less than $582,500, and about the remaining 50% sold for higher than that price.

   (b) The sample mean decreases, such that $\bar{x}$ = 610.5 (that is, $610500), and the sample median stays the same, such that $\tilde{x}$ = 582.5 (that is, $582500).

   (c) Deleting the smallest 2 observations (350, 408) and largest 2 observations (815, 1285), the trimmed mean $\bar{x}_{tr}$ is 591.2 ($591200).

   (d) The 15% trimmed mean $\bar{x}_{tr}$ is = (591.2+596.3)/2 = 593.7 ($593700). That is, the 15% trimmed mean is the average of the 20% and the 10% trimmed means.

3. (a) The sample mean will be larger than sample median.

```
 2   04566778
 3   012334466667
 4   4678
 5   3
 6
 7
 8
 9
10   1
```

   (b) Sample mean = 3.65, Sample median = 3.35.
   (c) Can be increased as much as one likes, but can't be made any smaller than 3.4 without affecting the median.

5. Sample mean will be larger than the sample median, because there is a potential outlier (59.31) which will pull the mean towards itself.
   Report the median, because that is resistant to outliers, and hence gives a better measure of typical energy use.

7. The sorted data are: 17, 29, 35, 48, 57, 79, 86, 92, 100+, 100+. The median is the average of the two middle observations: $\tilde{x}$ = (57+79)/2 = 68. On the other hand, the sample mean $\bar{x}$ can not be computed because the two largest values are not known (we only know that they are greater than 100). Several trimmed means can be computed: for example, by dropping the smallest and largest *two* observations (i.e., with 20% trimming), the trimmed mean is (35+48+…+86+92)/8 = 397/8 = 66.2. Similarly, the 30% trimmed mean is 67.5.

9. (a) $\mu = \int_0^2 x(.5x)dx = .5\left[\frac{x^3}{3}\right]_0^2 = 4/3$. The mean μ does not equal 1 because the density curve is not symmetric around x = 1.

(b) Half the area under the density curve to the left (or right) of the median $\tilde{\mu}$, so, $.50 = \int_0^{\tilde{\mu}} .5x\,dx = .5\left[\frac{x^2}{2}\right]_0^{\tilde{\mu}} =$ $(\tilde{\mu}^2)/4$, or, $\tilde{\mu}^2 = 4(.5) = 2$ and $\tilde{\mu} = 1.414$. $\mu < \tilde{\mu}$ because the density curve is negatively skewed.

(c) $\mu \pm \frac{1}{2} = \frac{4}{3} \pm \frac{1}{2} = \frac{5}{6}$ and $\frac{11}{6}$. The area under the curve between these two values is $\int_{5/6}^{11/6} .5x\,dx = \left[\frac{x^2}{4}\right]_{5/6}^{11/6}$ $= .667$. Similarly, the proportion of the times that are within one-half hour of $\tilde{\mu}$ is: $\int_{.914}^{1.914} .5x\,dx = \left[\frac{x^2}{4}\right]_{.914}^{1.914} =$ .707.

11. $\mu = \int_1^2 2x\left(1 - \frac{1}{x^2}\right)dx = \int_1^2 2x\,dx - \int_1^2 \frac{2}{x}dx = \left[x^2\right]_1^2 - \left[2\ln(x)\right]_1^2 = (2^2\text{-}1^2) - 2(\ln(2)\text{-}\ln(1)) = 3 - 2(.693147) =$ 1.61371, or about 1.614. For the median, $\tilde{\mu}$, half of the area under the density must lie to the left of $\tilde{\mu}$, so .5 = $\int_1^{\hat{\mu}} 2\left(1 - \frac{1}{x^2}\right)dx = \left[2x\right]_1^{\hat{\mu}} + \left[\frac{2}{x}\right]_1^{\hat{\mu}} = 2(\tilde{\mu}\text{-}1) + 2(1/\tilde{\mu} - 1)$. Solving for $\tilde{\mu}$ results in a quadratic equation $2\tilde{\mu}^2$ - $4.5\tilde{\mu} + 2 = 0$, whose roots are .61 and 1.64; only the root $\tilde{\mu} = 1.64$ is feasible (since it lies in the interval from 1 to 2). The proportion of x values that lie within the mean and median is $\int_{1.614}^{1.64} 2\left(1 - \frac{1}{x^2}\right)dx = \left[2x\right]_{1.614}^{1.64} + \left.\frac{2}{x}\right]_{1.614}^{1.64} =$ .032.

13. $\mu = \sum_{x=0}^{4} x \cdot p(x) = 0(.4) + 1(.1) + 2(.1) + 3(.1) + 4(.3) = 1.8$

Section 2.2

15. (a) $\bar{x} = \frac{1}{n}\sum_i x_i = 1939.4.$

Deviations from the mean:

Obsn.	2006.1	2065.2	2118.9	1686.6	1966.9	1792.5
Deviation from mean	66.7	125.8	179.5	-252.8	27.5	-146.9

(b) Sample variance, $s^2 = [66.7^2 + 125.8^2 + 179.5^2 + (-252.8)^2 + 27.5^2 + (-146.9)^2]/(6-1) = 27747.7$

Sample standard deviation = $\sqrt{27747.7} = 166.6$

(c) Done

17. (a) Group 1: $\bar{x} = \frac{1}{n}\sum_i x_i = 9.86;\ s=\sqrt{\dfrac{\sum x_i^2 - \dfrac{\left(\sum x_i\right)^2}{n}}{n-1}} = \sqrt{\dfrac{723-\dfrac{(69)^2}{7}}{7-1}} = 2.67$

Group 2: $\bar{x} = \frac{1}{n}\sum_i x_i = 8.93;\ s=\sqrt{\dfrac{\sum x_i^2 - \dfrac{\left(\sum x_i\right)^2}{n}}{n-1}} = \sqrt{\dfrac{591.75-\dfrac{(62.5)^2}{7}}{7-1}} = 2.37$

(b) Group 1: Range = Max – Min = 14 – 7 = 7;
Group 2: Range = Max – Min = 13 – 5 = 8

(c) Here's the dotplot

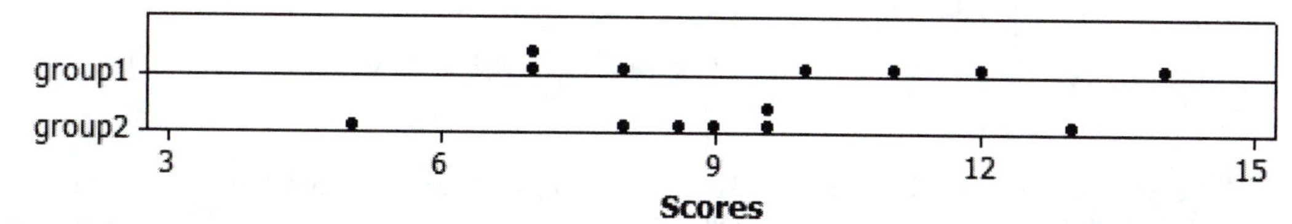

(d) Recall that standard deviation measures the distance between observations from the mean, whereas the range looks at the distance between the minimum and maximum. If in a dataset almost all the observations are close together but the minimum and the maximum are very far apart and very far away from the rest of the data (as happens for group 2), that group will have a smaller standard deviation (2.37) and larger range (8) compared to a dataset where the minimum and the maximum are not as extreme compared to the rest of the data (as happens in group 1, with standard deviation = 2.67, and range =7). Thus, a dataset where the min and max are outliers is likely to have higher range and lower standard deviation compared to a dataset where the min and max are not outliers.

19. The sample mean, $\bar{x} = \frac{1}{n}\sum x_i = 17.67\%$.

The sample variance, $s^2 = \frac{\sum (x_i - \bar{x})^2}{n-1} = \frac{91.64 + 7.99 + ... + 21.13}{9-1} = \frac{328.7516}{9-1} = 41.094$

Thus, sample standard deviation $= \sqrt{41.094} = 6.41\%$

In the sample, the average porosity of concrete cores was about 17.67%. And, the typical distance (difference) between a concrete core's porosity from the sample average was about 6.41%.

21. To somewhat simplify the algebra, begin by subtracting 76,000 from the original data. This transformation will affect each date value and the mean. It will not affect the standard deviation.

$x_1 = 683, \quad x_4 = 1{,}048, \quad \bar{y} = 831$

$n\bar{x} = (4)(831) = 3{,}324$ so, $x_1 + x_2 + x_3 + x_4 = 3{,}324$

and $x_2 + x_3 = 3{,}324 - x_1 - x_4 = 1{,}593$ and $x_3 = (1{,}593 - x_2)$

Next, $s^2 = (180)^2 = \left[\frac{\sum x_i^2 - \frac{(3324)^2}{4}}{3}\right]$

So, $\sum x_i^2 = 2{,}859{,}444$, $x_1^2 + x_2^2 + x_3^2 + x_4^2 = 2{,}859{,}444$ and $x_2^2 + x_3^2 = 2{,}859{,}444 - x_1^2 + x_4^2 = 1{,}294{,}651$

By substituting $x_3 = (1593 - x_2)$ we obtain the equation $x_2^2 + (1{,}593 - x_2)^2 - 1{,}294{,}651 = 0$.

$x_2^2 - 1{,}593x_2 + 621{,}499 = 0$

Evaluating for x_2 we obtain $x_2 = 682.8635$ and $x_3 = 1{,}593 - 682.8635 = 910.1365$. Thus, $x_2 = 76{,}683$ $x_3 = 76{,}910$.

23. Let X = the number of drivers who travel between a particular origin and destination during a designated time period. X has a Poisson distribution with $\lambda = 20$.

(a) $\mu_X = \lambda = 20$

Find $P(\mu - 5 \le x \le \mu + 5) = P(15 \le x \le 25)$

Using Table III with $\lambda = 20$, we obtain:

$P(15 \le x \le 25) = .052 + .065 + .076 + .084 + .089 + .089 + .085 + .077 +$
$.067 + .056 + + .045 = .785$

(b) $\sigma_x = \sqrt{\lambda} = \sqrt{20} = 4.47$

Find $P(\mu - \sigma \le x \le \mu + \sigma)$

$P(20 - 4.47 \le x \le 20 + 4.47) = P(15.53 \le x \le 24.47)$

But, since X is an integer–valued random variable, only the integers between 16 and 24 satisfy this requirement. So, we find $P(16 \le x \le 24)$ using Table III with $\lambda = 20$, we obtain:

$P(16 \le x \le 24) = .065 + .076 + .084 + .089 + + .089 + .085 + .077 + .067 + .056$
$= .688$

25. (a) $\sigma = \sqrt{\sum (x - \mu)^2 p(x)}$

$$\sigma = \sqrt{\begin{aligned}&(0-1.8)^2(.4) + (1-1.8)^2(.1) + (2-1.8)^2(.1) + \\ &(3-1.8)^2(.1) + (4-1.8)^2(.3)\end{aligned}}$$

$\sigma = \sqrt{2.96} = 1.72$

(b) $P(\mu - \sigma \le x \le \mu + \sigma) = P(.08 \le x \le 3.52) = P(1 \le x \le 3) = .3$

$P(x \rangle \mu + 3\sigma) + P(x \langle \mu - 3\sigma) = P(x \rangle 6.96) + P(x \langle -3.36) = 0 + 0 = 0$

27. Extending the result from Exercise 26, we know:

$$\sigma^2 = \int_{-\infty}^{\infty} (x - \mu)^2 f(x) dx = \left[\int_{-\infty}^{\infty} x^2 f(x) dx \right] - \mu^2$$

So, in the case of a uniform from a to b,

$$\sigma^2 = \left[\int_a^b x^2 \left(\frac{1}{b-a} \right) dx \right] - \mu^2 = \left(\frac{1}{b-a} \right) \left[\frac{x^3}{3} \right]_a^b - \mu^2$$

$$= \frac{1}{b-a} \left(\frac{b^3}{3} - \frac{a^3}{3} \right) - \left(\frac{a+b}{2} \right)^2 = \frac{1}{b-a} \left(\frac{1}{3} \right) \left(b^3 - a^3 \right) - \left(\frac{1}{4} \right) (a+b)^2$$

$$= \frac{1}{3} \left(a^2 + ab + b^2 \right) - \frac{1}{4} (a+b)^2 = \frac{1}{12} (b-a)^2$$

For the task completion time application, a = 4 and b = 6.

$$\mu = \left(\frac{a+b}{2}\right) = \left(\frac{4+6}{2}\right) = 5 \qquad \sigma^2 = \frac{(b-a)^2}{12} = \frac{(6-4)^2}{12} = \frac{1}{3}$$

$$P(x \rangle \mu+\sigma) + P(x \langle \mu-\sigma) = P\left(x \rangle 5+\sqrt{1/3}\right) + P\left(x \langle 5-\sqrt{1/3}\right)$$

$$= P(x \rangle 5.577) + P(x \langle 4.423) = \frac{6-5.577}{2} + \frac{4.423-4}{2} = .2115 + 2115 = .423$$

29. X is binomial with $\pi = .2$ and $n = 25$.

$\sigma^2 = n\pi(1-\pi) = (25)(.20)(.80) = 4 \quad \sigma = 2$

Also, $\mu = n\pi = (25)(.20) = 5$

$P(x \rangle \mu + 2\sigma) = P(x \rangle 5 + 2(2)) = P(x \rangle 9)$

Using Table II:

$P(x \rangle 9) = P(x \geq 10) = .011 + .004 + .002 = .017$

(Notice that $P(x \geq 13) \approx 0$)

31. $$\int_{\mu+\sigma}^{\infty} \lambda e^{-\lambda x} dx = \left[-e^{-\lambda x}\right]_{\mu+\sigma}^{\infty} = 0 - \left(-e^{-\lambda(\mu+\sigma)}\right) = e^{-\lambda(\mu+\sigma)}$$

We need μ and σ for an exponential random variable.

$$\mu = \int_0^{\infty} x\lambda e^{-\lambda x} dx$$

Using integration by parts let $u = x,\ dv = \lambda e^{-\lambda x} dx,\ du = dx,\ v = -e^{-\lambda x}$

So, $\int_0^{\infty} x\lambda e^{-\lambda x} dx = -xe^{-\lambda x}\Big]_0^{\infty} - \int_0^{\infty} -e^{-\lambda x} dx = 0 + \left[\frac{-1}{\lambda} e^{-\lambda x}\right]_0^{\infty} = \left(\frac{1}{\lambda}\right)$

That is, $\mu = \left(\frac{1}{\lambda}\right)$.

Now, $\sigma^2 = \left[\int_0^{\infty} x^2 \lambda e^{-\lambda x} dx\right] - \mu^2$

Using integration by parts twice:

Fist we obtain: $u = x^2,\ dv = \lambda e^{-\lambda x} dx,\ du = 2xdx,\ v = -e^{-\lambda x}$

$$\int_0^\infty x^2 \lambda e^{-\lambda x}dx = \left[-x^2 e^{-\lambda x}\right]_0^\infty - \int -2xe^{-\lambda x}dx = 0 + 2\int xe^{-\lambda x}dx$$

Then, using integration by parts again: $u = x,\quad dv = e^{-\lambda x}dx,\quad du = dx,\quad v = -\frac{1}{\lambda}e^{-\lambda x}$

$$2\int xe^{-\lambda x}dx = 2\left[\left(\frac{xe^{-\lambda x}}{\lambda}\right)_0^\infty - \int_0^\infty -\frac{1}{\lambda}e^{-\lambda x}dx\right] = 2\left(0 - \left[\frac{e^{-\lambda x}}{\lambda^2}\right]_0^\infty\right) = 2\left(\frac{1}{\lambda^2}\right) = \frac{2}{\lambda^2}$$ Thus,

$$\sigma^2 = \frac{2}{\lambda^2} - \mu^2 = \left(\frac{2}{\lambda^2}\right) - \left(\frac{1}{\lambda}\right)^2 = \left(\frac{1}{\lambda^2}\right)$$

That is, $\sigma^2 = \left(\frac{1}{\lambda^2}\right)$ and $\sigma = \left(\frac{1}{\lambda}\right)$

Then, finally: $\int_{\mu+\sigma}^\infty \lambda e^{-\lambda x}dx = e^{-\lambda(\mu+\lambda)} = e^{-\lambda\left(\frac{1}{\lambda}+\frac{1}{\lambda}\right)} = e^{-2} = .135$

Section 2.3

33. (a) The lower half of the data set: 108 120 122 122 126 127 133, whose median, and therefore the lower quartile, is 122. The top half of the data set: 133 134 135 135 137 142 154, whose median, and therefore the upper quartile, is 135. So, the IQR = (135-122) = 13.

(b) A boxplot (created in Minitab) of this data appears below; there is a negative skew to the data. The variation seems quite large. There are no outliers.

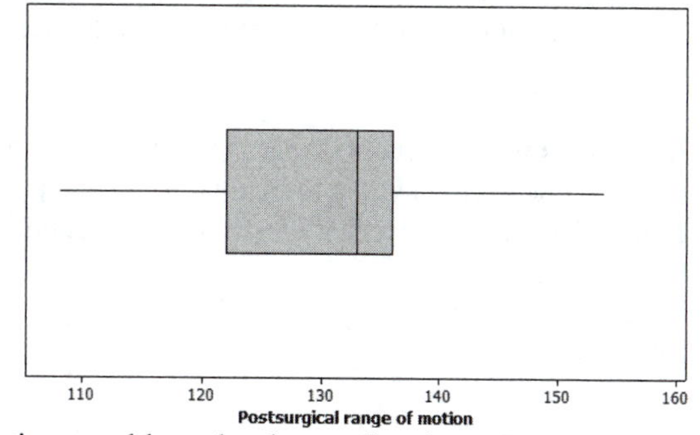

(c) An observation would need to be smaller than 122 – 1.5(13) = 102.5, or larger than 136+1.5(13) = 154.5 to be classified as a mild outlier.

An extreme outlier would need to be smaller than 122 – 3(13) = 83, or larger than 136+3(13) = 174. Since the minimum and maximum observations in the data are 108 and 154 respectively, we conclude that there are no outliers, of either type, in this data set.

(d) Any change to the maximum will NOT change IQR at all. Make the maximum 134 and the IQR is still 13. Change it to anything you want and the IQR is still 13. This is due to the duplicate values of 122 and 135 that appear.

35. Since 5% of all lengths exceed 3.75 μm, then 3.75 is the 95th percentile of the distribution. Because the circuits are normally distributed, 3.75 is 1.645 standard deviations above the mean; that is, $3.75 = \mu + 1.645\sigma$. Furthermore, 3.85 is the 99th percentile of the distribution, and so it is 2.33 standard deviations above the mean: $3.85 = \mu + 2.33\sigma$. We then have two equations and two unknowns:

$$\begin{aligned} \mu + 1.645\sigma &= 3.75 \\ \mu + 2.33\sigma &= 3.85 \end{aligned}$$

Subtracting the top equation from the bottom equation yields $.685\sigma = .10$, and so $\sigma = .10/.685 = .146$. Then substituting $\sigma = .146$ into either of the equations gives $\mu = 3.51$.

37. A boxplot (created in Minitab) of this data appears below.

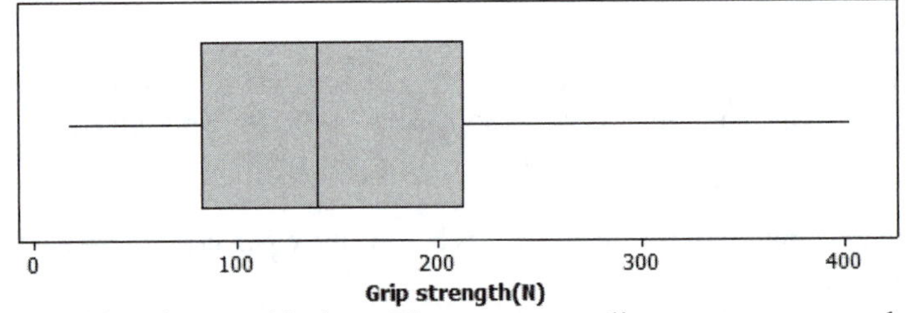

There is a positive skew to this data. There are no outliers, extreme or moderate.

39. The most noticeable feature of the comparative boxplots is that machine 2's sample values have considerably more variation than do machine 1's sample values. However, a typical value, as measured by the median, seems to be about the same for the two machines. The only outlier that exists is from machine 1.

41. A comparative boxplot (created in Minitab) appears below. Farm homes appear to have lower median values but higher IQR values for endotoxin concentration in dust from vacuum cleaners compared to urban homes (based on the sample data). There was one urban home and one farm home that were outliers in the respective samples of homes.

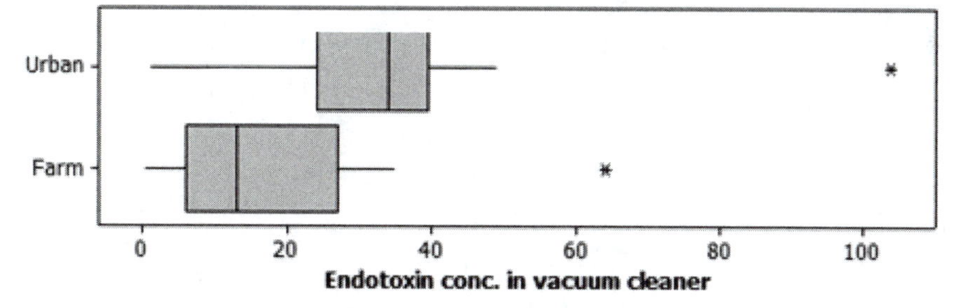

43. X= fracture strength of silicon nitride braze joints. X follows a Weibull distribution with $\alpha=5$ $\beta=125$

(a) The lower quartile of X: $.25=1-e^{-(q_\ell/125)^5}$

$$e^{-(q_\ell/125)^5}=.75$$

$$-\left(q_\ell/125\right)^5=\ell n(.75)$$

$$q_\ell=125\left((-\ell n(.75))^{1/5}\right)=97.43$$

The upper quartile of X is: $.75=1-e^{-\left(q_u/125\right)^5}$

$$q_u=125\left((-\ell n(.25))^{1/5}\right)=133.44$$

$$IQR=(q_u-q_l)=(133.44-97.43)=36.01$$

(b) $q_l=12.5\left((-\ell n(.75))^{1/5}\right)=9.74$

$$q_u=12.5\left((-\ell n(.25))^{1/5}\right)=13.34$$

$$IQR=(13.34-9.74)=3.6$$

Section 2.4

45. There is some wiggling in the plot. There are also some small gaps. However, the general pattern is reasonably straight and a departure from linearity is not clear-cut. One should not rule out normality of the tension distribution.

47. The quantile plot (created using the same technique as outlined in the textbook) is shown below:

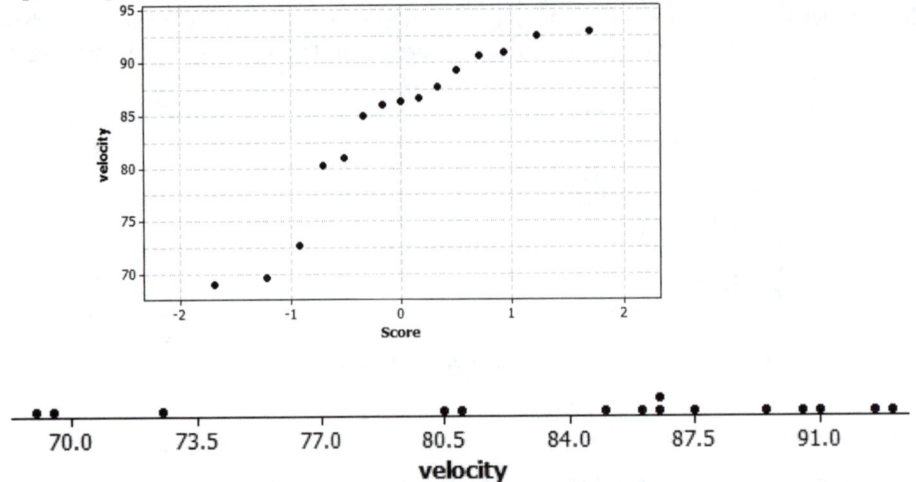

No, it does not appear plausible that data come from a normal distribution, because the pattern in the normal quantile plot does not look linear.

49. This Weibull quantile plot was created in Minitab.

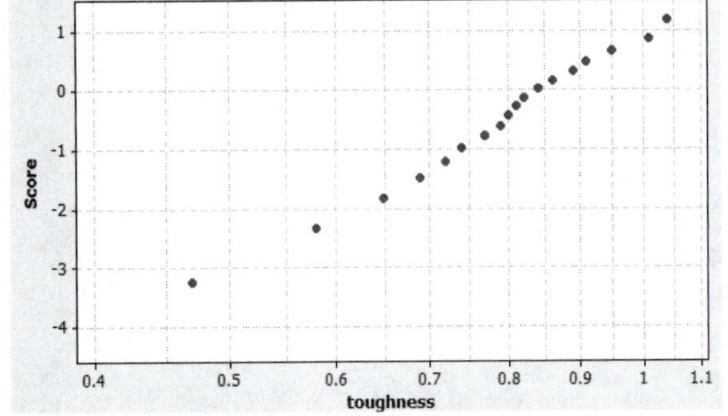

Appears plausible that the data on toughness comes from a Weibull distribution because the pattern in the Weibull quantile plot looks fairly linear.

51. Clearly, the variable IDT is not normally distributed, since it's normal quantile plot is nonlinear. IDT is likely to be lognormally distributed since the normal quantile plot of $\ell n(ITD)$ is quite linear.

53. (a) A normal quantile plot of x follows.
Clearly the variable, hourly median power, is not normally distributed, as the normal quantile plot is curvilinear.

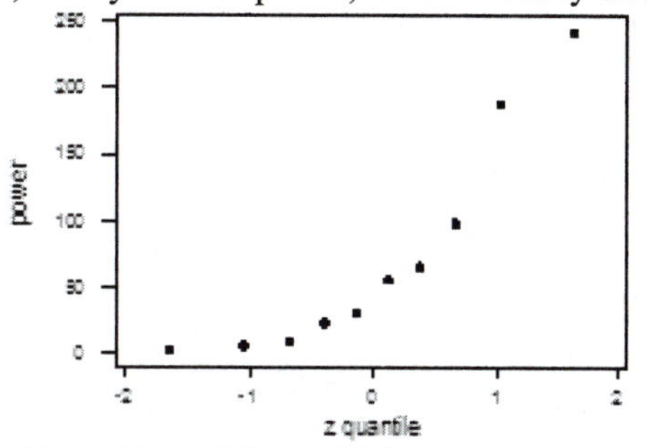

(b) By taking the natural logarithm of the variable and constructing a normal quantile plot of it, we obtain:
This plot looks quite linear indicating that it is plausible that these observations were sampled from a lognormal distribution.

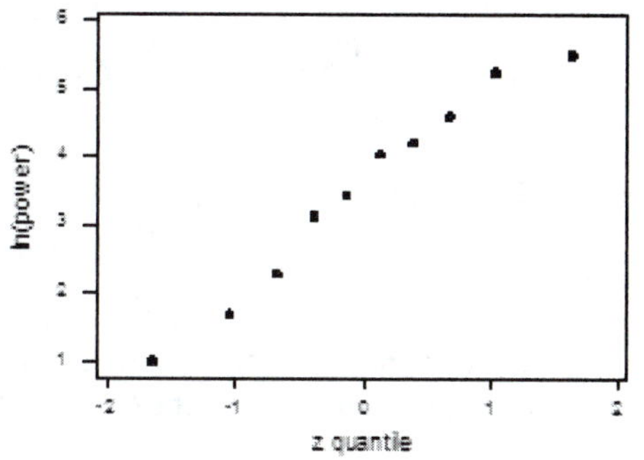

Supplementary Exercises

55. Answers may vary; May include dotplots, stem and leaf plots, boxplots or histograms, along with mean and SD, or the 5-number summary, or both. For example, we can see from the boxplot that the distribution is skewed right.

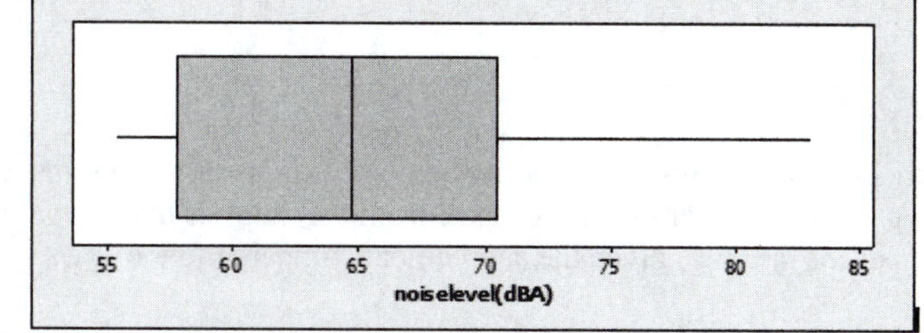

From Minitab:

Descriptive Statistics: noiselevel(dBA)

Variable	N	Mean	StDev	Variance	Minimum	Q1	Median	Q3	Maximum	IQR
noiselevel(dBA)	77	64.887	7.803	60.882	55.300	57.800	64.700	70.400	83.000	12.600

57. (a) $ks_{k+1}^2 = \sum_{i=1}^{k+1}(x_i - \overline{x}_{k+1})^2 = \sum_{i=1}^{k+1} x_i^2 - (k+1)\overline{x}_{k+1}^2$

$$= \sum_{i=1}^{k} x_i^2 - k\overline{x}_k^2 + x_{k+1}^2 + k\overline{x}_k^2 - (k+1)\overline{x}_{k+1}^2$$

$$= (k-1)s_k^2 + \{x_{k+1}^2 + k\overline{x}_k^2 - (k+1)\overline{x}_{k+1}^2\}$$

When substituting $\overline{x}_{k+1} = \left(\frac{k\overline{x} + x_{k+1}}{k+1}\right)$, the expression in braces simplifies to: $\frac{k(x_{k+1} - \overline{x}_k)^2}{(k+1)}$. Thus:

$$ks_{k+1}^2 = (k-1)s_k^2 + \frac{k}{k+1}(x_{k+1} - \overline{x}_k)^2$$

(b) k = 15; $\overline{x}_{15} = 12.58$; $s_{15} = .512$; $x_{16} = 11.8$ $\overline{x}_{16} = \left(\frac{(15)(12.58) + 11.8}{16}\right) = 12.53125$;

$$(15)s_{16}^2 = (14)(.512)^2 + \left(\frac{15}{16}\right)(11.8 - 12.58)^2 = 4.240391;\ s_{16} = .532$$

59. (a) Answers may vary. May include dotplots, stem and leaf plots, boxplots or histograms, along with mean and SD, or the 5-number summary, or both. For example, we can see from the dotplots that the distributions are similar except for two outliers in the control group.

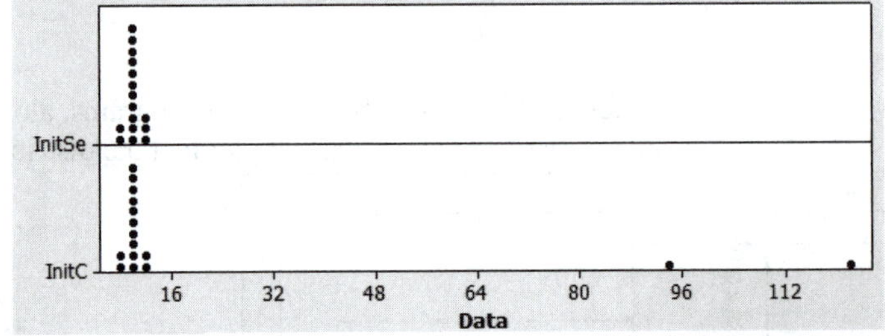

(b) Answers may vary. May include dotplots, stem and leaf plots, boxplots or histograms, along with mean and SD, or the 5-number summary, or both. For example, we can see from the dotplots that the values of milk Se concentration are much higher for the cows given the supplement compared to the cows in the control group.

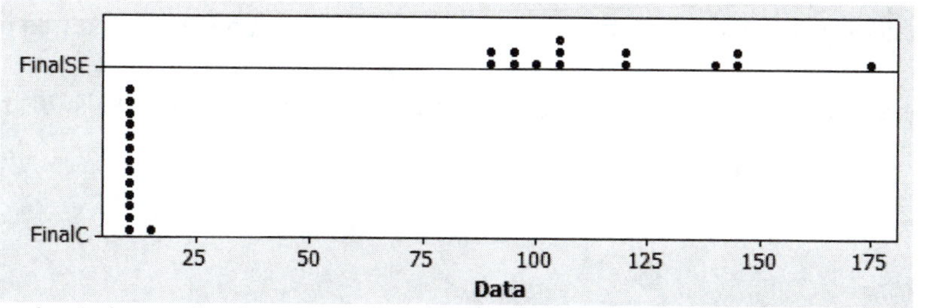

61. (a) Chebyshev's inequality is more conservative than is the empirical rule. This is because Chebyshev's rule makes no assumptions about the shape of the distribution of the values while the empirical rule assumes the shape of the distribution in approximately normal.

	Chebyshev's Rule	**Empirical Rule**
Percentage within 1 standard deviation	No statement	Approximately 68%
Percentage within 2 standard deviations	At least 75%	Approximately 95%
Percentage within 3 standard deviations	At least 89%	Approximately 99.7%

(b) $\mu = 100 \Rightarrow \lambda = .01$

$\sigma = 100$

$$\mu \pm \sigma \Rightarrow 100 \pm 100 \Rightarrow (0,\ 200)\ ;\ \int_0^{200} .01e^{-.01x}dx = 1 - e^{-.01(200)} = .8647$$

$$\mu \pm 2\sigma \Rightarrow 100 \pm 200 \Rightarrow (0,\ 300)\ ;\ \int_0^{300} .01e^{-.01x}dx = 1 - e^{-.01(300)} = .9502$$

$$\mu \pm 3\sigma \Rightarrow 100 \pm 300 \Rightarrow (0,\ 400)\ ;\ \int_0^{400} .01e^{-.01x}dx = 1 - e^{-.01(400)} = .9817$$

So:

Percentage within	**Chebyshev's Rule**	**Exponential**
1σ	No statement	86.47%
2σ	At least 75%	95.02%
3σ	At least 89%	98.17%

(c) In part (a) a normal distribution is assumed, since the empirical rule is applied.

In part (b) an exponential distribution with $\lambda = .01$ is specified, a positively skewed distribution.

In either case Chebyshev's inequality may not accurately estimate any particular distribution as it must accommodate all distributions.

63. (a) The ordered values are: 7.9 8.5 8.8 9.2 9.3 9.6 9.8 10.5 10.7 11.0 12.1 12.2 13.2 13.7 16.6

By trimming the smallest and largest values and averaging the remaining 13 values a $\left(100\left(\frac{1}{15}\right)\%\right) = 6.7\%$ trimmed mean is produced. $\bar{x}_{tr(6.7)} = 10.67$

By trimming the two smallest and two largest values and averaging the remaining 11 values a $\left(100\left(\frac{2}{15}\right)\%\right) =$ 13.3% trimmed mean is produced.

$\bar{x}_{tr(13.3)} = 10.58$

(b) Perhaps by averaging the 6.7% and 13.3% trimmed means since $\left(\frac{(6.7\% + 13.3\%)}{2}\right) = 10\%$. Thus:

$$\bar{x}_{tr(10)} = \left(\frac{(10.67 + 10.58)}{2}\right) = 10.625$$

(c) First compute $\bar{x}_{tr(6.25)}$. Then compute $\bar{x}_{tr(12.5)}$.

Finally, interpolate between these two trimmed mean values to obtain $\bar{x}_{tr(10)}$.

65. The measures that are sensitive to outliers are the mean and the midrange. The mean is sensitive because all values are used in computing it. The midrange is sensitive because it uses only the most extreme values in its computation.

The median, the trimmed mean, and the midhinge are not sensitive to outliers.

The median is the most resistant to outliers because it uses only the middle value (or values) in its computation.

The trimmed mean is somewhat resistant to outliers. The larger the trimming percentage, the more resistant the trimmed mean becomes.

The midhinge, which uses the quartiles, is reasonably resistant to outliers because both quartiles are resistant to outliers.

67. (a) Answers may vary. May include dotplots, stem and leaf plots, boxplots or histograms, along with mean and SD, or the 5-number summary, or both. For example, we can see from the summary statistics and the dotplots that the values of aortic root diameter tend to be higher, on average, for males compared to females, but there is more variability among the females compared to the males.

Descriptive Statistics: Male, Female

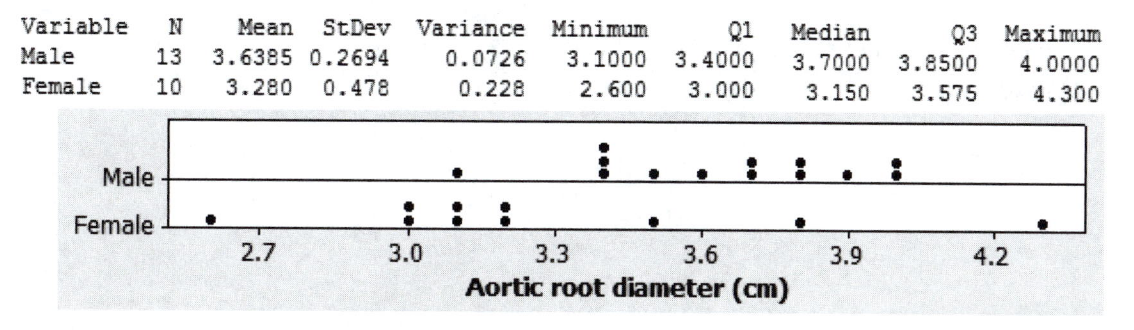

Variable	N	Mean	StDev	Variance	Minimum	Q1	Median	Q3	Maximum
Male	13	3.6385	0.2694	0.0726	3.1000	3.4000	3.7000	3.8500	4.0000
Female	10	3.280	0.478	0.228	2.600	3.000	3.150	3.575	4.300

(b) For females, 3.7cm, and for males 3.2cm

69. $\begin{pmatrix} \textit{proportion of } x \textit{ values} \\ \textit{satisfying } x<5.33 \end{pmatrix} = \begin{pmatrix} \textit{proportion of } z \textit{ values} \\ \textit{satisfying } z<2.33 \end{pmatrix} = .99$

$\Rightarrow 2.33 = \left(\frac{5.33-\mu}{\sigma}\right)$

$\Rightarrow \mu = 5.33 - 2.33\sigma$

Also: $\begin{pmatrix} \textit{proportion of } x \textit{ values} \\ \textit{satisfying } x<1.72 \end{pmatrix} = \begin{pmatrix} \textit{proportin of } z \textit{ values} \\ \textit{satisfying } z<-1.28 \end{pmatrix} = .10$

$\Rightarrow -1.28 = \frac{1.72-\mu}{\sigma}$

$\Rightarrow \sigma = \left(\frac{1.72-\sigma}{-1.28}\right)$

So: $\mu = 5.33 - 2.33\left(\frac{1.72-\mu}{-1.28}\right) = 5.33 + 1.82(1.72-\mu)$

$\mu = 5.33 + [3.13 - 1.82\mu]$

$2.82\mu = 8.46$

$\mu = 3$ and $\sigma = \left(\frac{1.72-3}{-1.28}\right) = 1$

X is normally distributed with $\mu = 3$ and $\sigma = 1$.

Finally, $\begin{pmatrix} \textit{proportion of } x \textit{ values} \\ \textit{satisfying } x > 5 \end{pmatrix} = \begin{pmatrix} \textit{proportion of } z \textit{ values} \\ \textit{satisfying } z > 2 \end{pmatrix}$

= 1 - .9772 = .0228

$$\begin{pmatrix} \textit{proportion of } x \textit{ values} \\ \textit{satisfying } x < 2 \end{pmatrix} = \begin{pmatrix} \textit{proportion of } z \textit{ values} \\ \textit{satisfying } z < -1 \end{pmatrix} = .1587$$

Chapter 3
Bivariate and Multivariate Data and Distributions

Section 3.1

1. The scatterplot below was created on Minitab. There appears to be a negative, strong, fairly linear association between solar absorptance and temperature. Yes, the desired inverse relationship is visible in the graph, because as the temperature increases the solar absorptance tends to decrease.

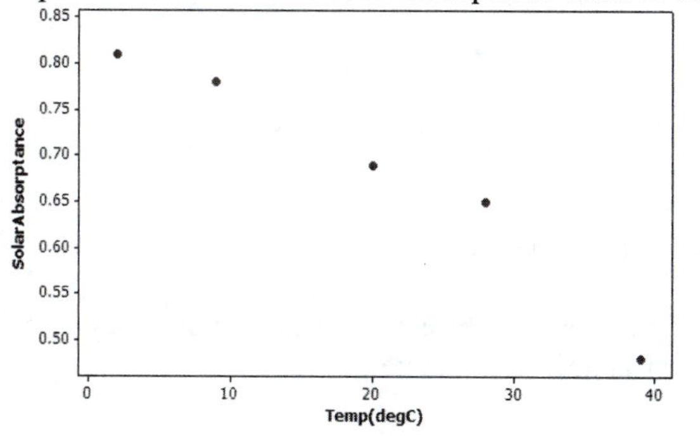

3. A scatter plot of the data appears below. There appears to be a strong positive linear relationship between the number of beds in a barrack and the cost of the building, except for one barrack where the cost of the building is unusually high relative to the number of beds in the barrack.

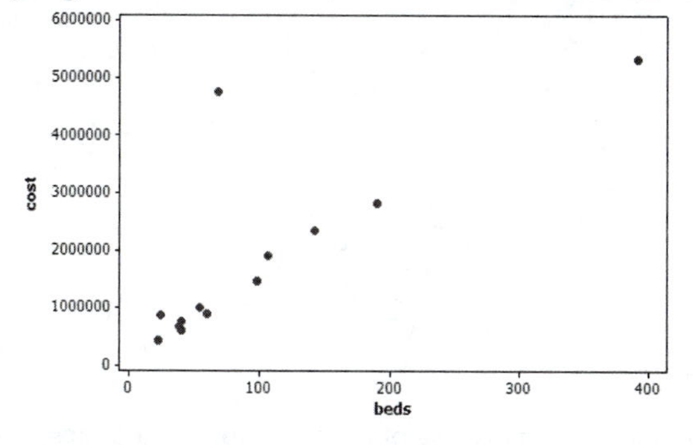

5. (a) The scatter plot with axes intersecting at (0,0) is shown below.

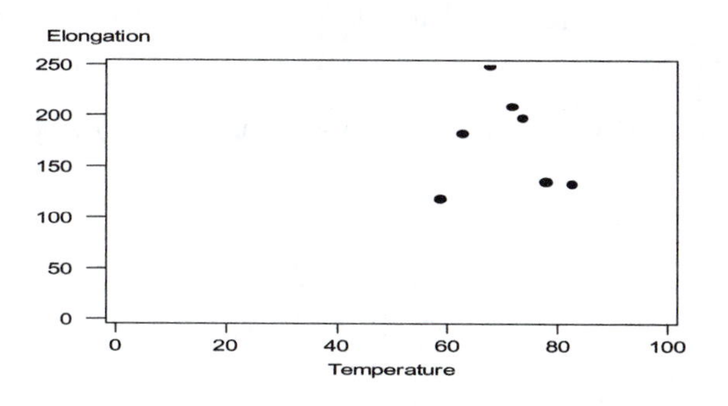

(b) The scatter plot with axes intersecting at (55,100) appears below. The plot in (b) makes it somewhat easier to see the nature of the relationship between the two variables.

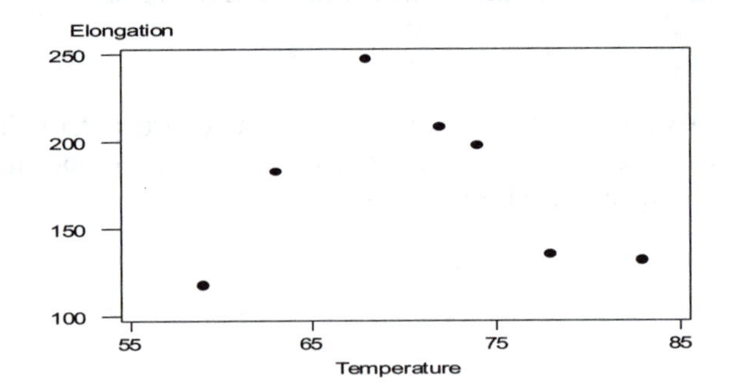

(c) A parabola appears to provide a good fit to both graphs.

7. The scatter plot of this data is shown below. There appears to be a negative, moderately strong, fairly linear relationship between alpha and average long-term annual temperature.

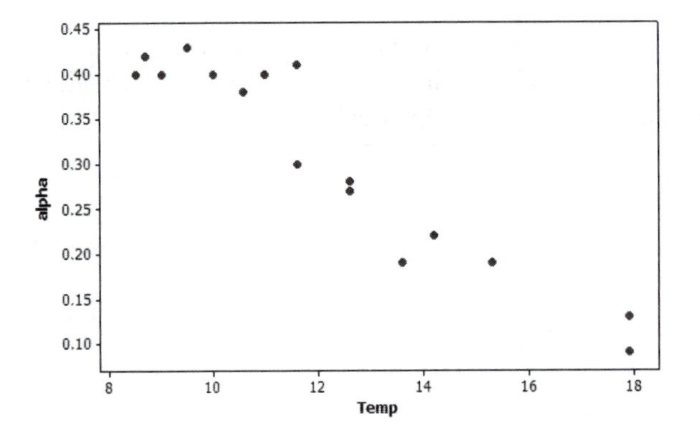

Section 3.2

9. (a) Positive - As temperature increases, people make more use of air conditioners and fans, which will increase utility costs.

(b) Negative - Higher interest rates makes loans more difficult to obtain and to pay off, which generally reduces the number of people seeking loans.

(c) Positive - Married couples tend to have similar educational backgrounds and educational level is positively correlated with income level.

(d) Over the entire range of speeds, there is little correlation: at very slow speeds, an increase in speed will improve fuel efficiency; at high speeds, additional increases in speed tend to reduce fuel efficiency.

(e) Negative - Gasoline costs decrease as fuel efficiency increases.

(f) Distance and GPA should be unrelated, so there should be little or no correlation.

11.
$$r = \frac{44185.87}{\sqrt{64732.83}\sqrt{130566.96}} = .481$$

Sample correlation coefficient = 0.481, which provides some evidence of a somewhat weak positive linear association between lead hip internal peak rotational velocity (x) and trailing hip peak external rotational velocity (y).

13. Most people acquire a license as soon as they become eligible. If for, example, the minimum age for obtaining a license is 16, then the time since acquiring a license, y, is usually related to age by the equation $y \approx x - 16$, which is the equation of a straight line. In other words, the majority of people in a sample will have y values that closely follow the line $y = x-16$.

15. Let d_0 denote the (fixed) length of the stretch of highway. Then, d_0 = distance = (rate)(time) = xy. Dividing both sides by x, gives the equation $y = d_0/x$ which means the relationship between x and y is curvilinear (in particular, the curve is a hyperbola). However, for values of x that are fairly close to one another, sections of this hyperbola can be approximated very well by a straight line with a negative slope (to see this, draw a picture of the function d_0/x for a particular value of d_0). This means that r should be closer to -.9 than to any of the other choices.

17. (a) $SS_{xx} = 37695 - (561)^2/9 = 2726$, $SS_{yy} = 40223 - (589)^2/9 = 1676.222$, and $SS_{xy} = 38281 - (561)(589)/9 = 1566.666$, so $r = \dfrac{1566.667}{\sqrt{2726}\sqrt{1676.222}} = .733$.

(b) $\bar{x}_1 = (70+72+94)/3 = 78.667$, $\bar{y}_1 = (60+83+85)/3 = 76$.

$\bar{x}_2 = (80+60+55)/3 = 65$, $\bar{y}_2 = (72+74+58)/3 = 68$.

$\bar{x}_3 = (45+50+35)/3 = 43.333$, $\bar{y}_3 = (63+40+54)/3 = 52.333$.

$S_{xx} = [(78.667)^2+(65)^2+(43.333)^2 - (78.667+65+43.333)^2/3] = 634.913$,

$S_{yy} = [(76)^2+(68)^2+(52.333)^2-(76+68+52.333)^2/3] = 289.923$,

$S_{xy} = [(78.667)(76)+(65)(68)+(43.333)(52.333)-(187)(196.333)/3] = 428.348$, so

$$r = \frac{428.348}{\sqrt{634.913}\sqrt{289.923}} = .9984.$$

(c) The correlation among the averages is noticeably higher than the correlation among the raw scores, so these points fall much closer to a straight line than do the unaveraged scores. The reason for this is that averaging tends to reduce the variation in data, making it more likely that the averages will fall close to a straight line than the more variable raw data.

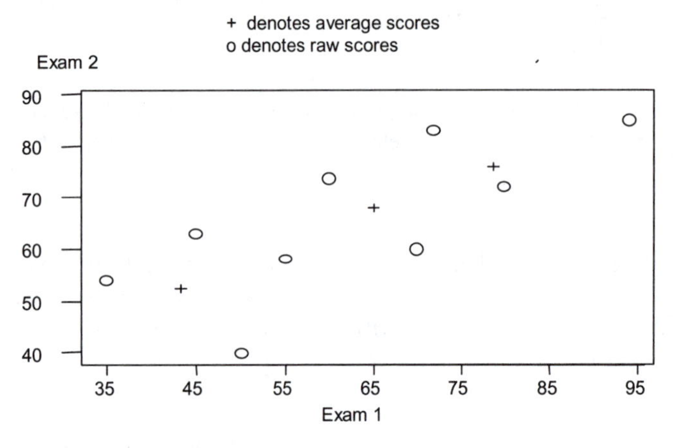

Section 3.3

19. (a) To obtain the least squares regression equation, first we will compute the slope and the vertical intercept using the equations provided in Section 3.3.

$$S_{xy} = \left[189482 - \frac{(756)(2944)}{15}\right] = 41104.4; \quad S_{xx} = \left[42228 - \frac{(756)^2}{15}\right] = 4125.6$$

$$\text{So, } b = \left(\frac{S_{xy}}{S_{xx}}\right) = \left(\frac{41104.4}{4125.6}\right) = 9.963$$

$$\text{Also: } a = \bar{y} - b\bar{x} = \left(\frac{2944}{15}\right) - (9.963)\left(\frac{756}{15}\right) = -305.881$$

Thus, the equation for the least squares line is:

$$\hat{y} = -305.881 + 9.963x$$

We need to compute r^2.

$$SSTo = S_{yy} = 3310341;$$

$$SSResid = SSTo - bS_{xy} = 3310341 - (9.963)(41104.4) = 2900807$$

$$So,\ r^2 = \left(1 - \frac{SSResid}{SSTo}\right) = 1 - \left(\frac{2900807}{3310341}\right) = 0.124$$

Thus, 12.4% of the observed variation in colony density can be explained by the approximate linear relationship between colony density and rock surface area.

(b) After deleting the second observation, the new least squares line becomes:

$$\hat{y} = 34.37 + 0.78x$$

The new $r^2 = 2.4\%$.

So, you can see the large effect the second observation had on the analysis. The estimate of the slope was decreased and the fit of the least squares line, as measured by r^2, is worse than it was.

21. (a) There appears to be a positive, moderately strong, linear relationship between axial strength and cube strength.

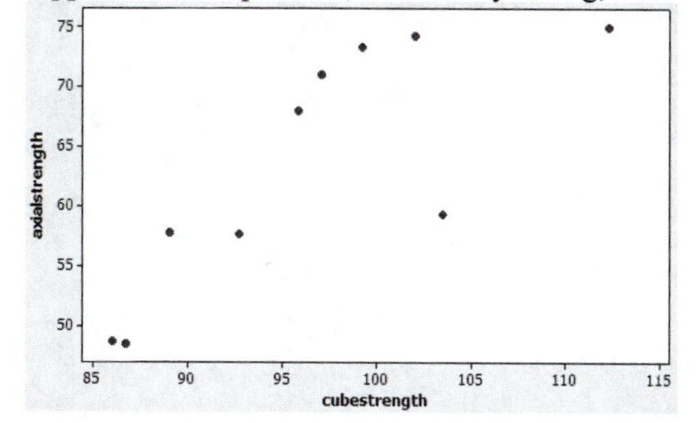

(b) Using Minitab to run this regression, we obtain the equation of the least squares line:

$\hat{y} = -31.80 + 0.987x$, where y = axial strength and x = cube strength

The increase in mean axial strength associated with a 1MPa increase in cube strength is estimated to be 0.987 MPa.

(c) r^2= 63.0%. That is, 63% of the observed variation in axial strength is attributable to the linear relationship between axial strength and cube strength.

(d) This is asking for the standard deviation about the least squares line = $\sqrt{\frac{SS\,\text{Re}\,sid}{n-2}} = 6.245$ from the Minitab output.

23. (a) Using Minitab to run this regression, we obtain the equation of the least squares line:

$\hat{y} = 11.013 - 0.448x$, where y = compressive strength and x = fiber weight

The decrease in mean compressive strength associated with a 1percenatge point increase in fibre weight is estimated to be 0.448 MPa.

(b) Using Minitab, coefficient of determination, r^2 = 69.4%

(c) When x = 6.5, Estimated compressive strength = 8.101MPa.

(d) No, because 25% is outside the range of the observed fibre weight values; using the estimated least squares line to make this prediction would be extrapolation.

When, x = 25, Estimated compressive strength = -.187MPa which is not meaningful since it is a negative value.

25. (a) The scatterplot was created on Minitab.

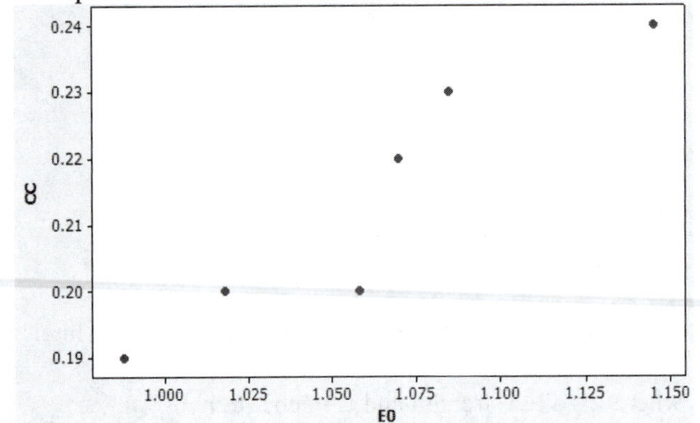

From the scatterplot, there appears to be a positive association between C_c and e_0, but the value of C_c doesn't appear to always go up when e_0 goes up, and by exactly the same amount always. Thus, it doesn't appear that the values of C_c and the values of e_0 have a perfect linear relationship.

(b) Using Minitab,

$\hat{y} = -0.01438 + 0.3367x$, where $y = C_c$, and $x = e_0$

(c) From Minitab, the coefficient of determination, $r^2 = 87.4\%$

(d) When $e_0 = 1.10$, Estimated compression index(C_c) = 0.22657

No, using the above least squares regression line for predicting compression index(C_c) when $e_0 = .8$ would be extrapolation, because this would assume that the same relationship holds true outside the observed range of data.

27. Data set #1

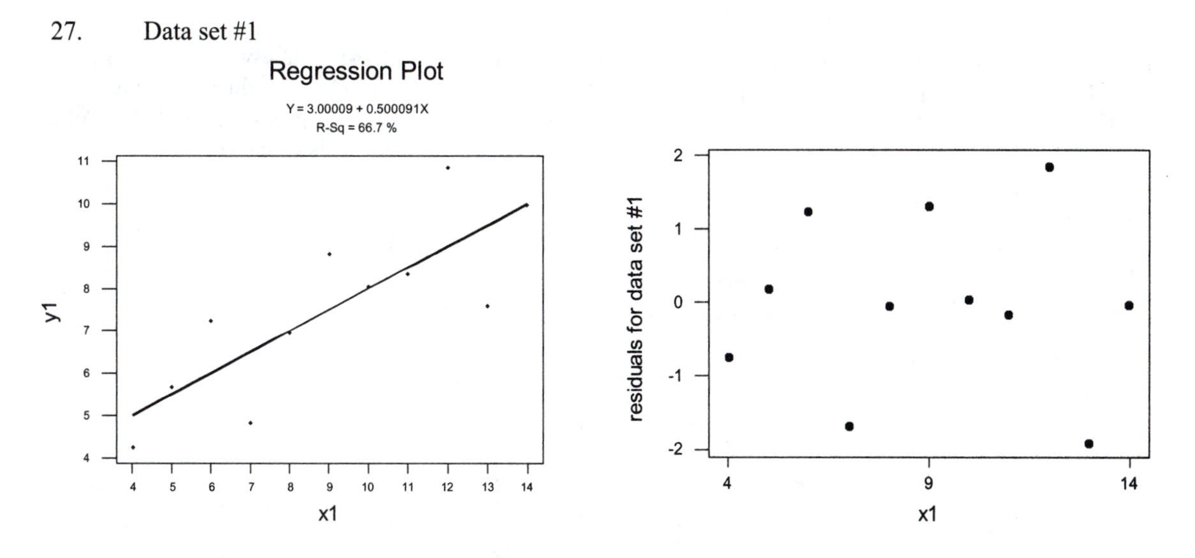

Fitting a straight line seems appropriate here. There is no indication of a problem. There is a fair amount of scatter around the least squares line, however. This fact is quantified by the r^2 value of about 67%.

Data set #2

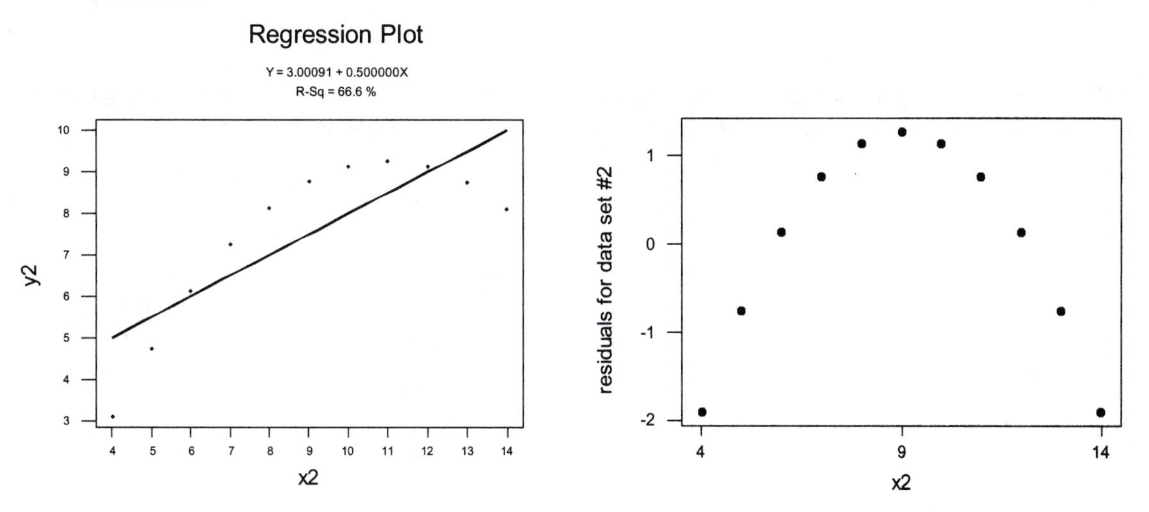

It is inappropriate to fit a straight line through this data. There is a quadratic relationship between these two variables. So, a model that incorporates this quadratic relationship should be fit instead of a linear fit.
Data set #3

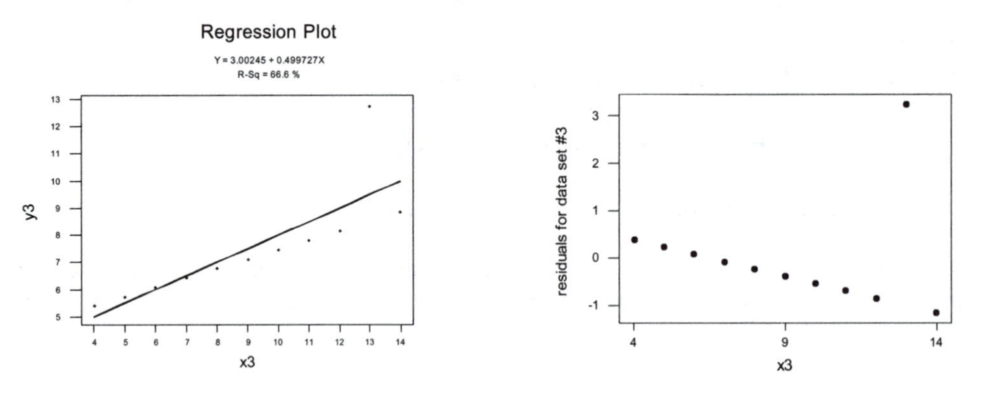

In this data set there is one pair of values that is a clear outlier (13.0, 12.74). The y-value of 12.74 is much larger than one would expect based on the other data. This pair of values exerts a lot of influence on the linear fit. It should be investigated. Perhaps it was a recording error or some other similar problem. With the outlier included, it is inappropriate to fit a straight line. With it excluded, an excellent linear fit is achieved.

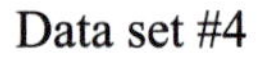

Data set #4

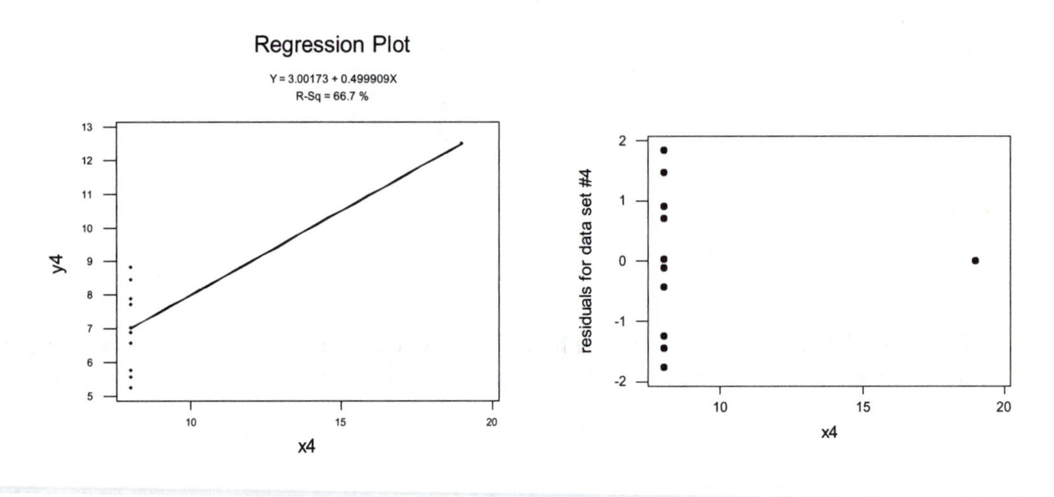

It is inappropriate to fit a straight line through this data. There is no linear relationship between the variables. There is one pair of observations that is a clear outlier (19.0, 12.50). It should be investigated.

Section 3.4

29. (a) No, scatterplot of P_{gt} versus Wt* shows a non linear association.
(b) Answers may vary. One possible transformation is a natural log transformation of the response variable, that is, Wt*. Using Minitab, we can obtain the following equation:

$$\hat{y} = 42.1 - 0.493x$$, where y = log(time), and x = Load

Which can be untransformed to $\hat{y}' = e^{42.1-0.493x}$, where y' = time, and x = Load

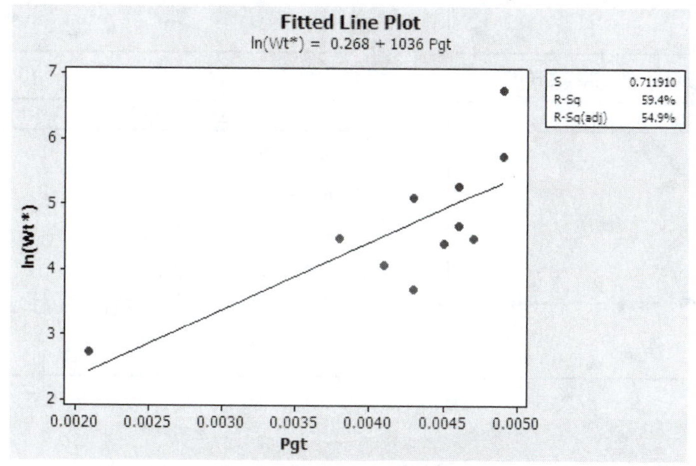

31. (a) Here are the scatterplots created in Minitab.

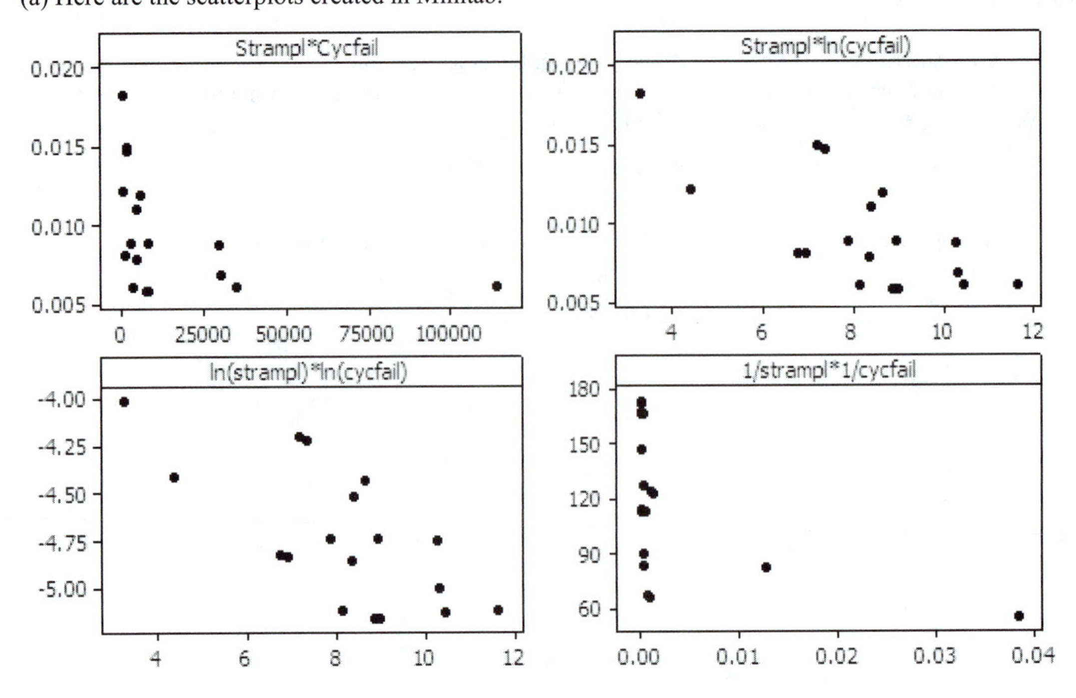

(b) The scatterplot of y versus ln(x) most closely resembles a linear association. Also, of the four possibilities, this relationship has the highest $r^2 = 0.497$.

(c) Using Minitab, we get the following output:

Regression Analysis: Strampl versus ln(cycfail)

```
The regression equation is Strampl = 0.0197 - 0.00128 ln(cycfail)
```

Thus, when Cycfail = 5000 ➔ Predicted Strampl = 0.0197 – 0.00128 ln(5000) = 0.0088

33. (a) It is not possible to transform this data as described in Section 3.4 so as to approximate a linear relationship between the transformed data. There is a quadratic relationship between yield and time to harvest, so a quadratic model should be fit.

(b) A quadratic fit to this data using Minitab produced:

$\hat{y} = 14.521 + .0432x - .00006x^2$.

At x=500, $\hat{y} = 21.121$. The residual plot shows no unusual pattern and R^2 =.780. The quadratic fit seems adequate.

Section 3.5

35. (a) $\hat{y}=0.0558+0.3749(11.5)+0.0028(40)=4.479$

We would predict a deposition rate of 4.479.
The corresponding residual is:
$(y-\hat{y})=(4.454-4.479)=-0.2025$

(b) SSResid = 0.03836
SSTo = 5.1109

$$R^2=1-\left(\frac{SSResid}{SSTo}\right)=1-\left(\frac{0.03836}{5.1109}\right)=.9925 \; or \; 99.25\%$$

99.25% of the observed variation in deposition rate can be attributed to the stated approximate relationship between deposition rate and the two predictor variables.

37. (a) The coefficient of multiple determination is R^2. $R^2 = 92.34\%$ in this regression.

92.34% of the observed variability in hydrocarbon deposition can be attributed to the given multiple regression model involving x_1 and x_2.

(b) $x_1 = 20{,}000$ $x_2 = 0.001$
$\hat{y}=-33.46+0.00206(20{,}000)+29836(0.001)=37.576$

(c) Yes, it is legitimate to interpret b_2 in this way, because when x_1 is held constant, the relationship between y and x_2 is linear, and then b_2 can be interpreted as the same way as a slope of a straight line.

39. (a) Here are the various fitted functions, and the corresponding coefficient of multiple determination, r^2:

Fitted function	r^2
$\hat{y}=a+b_1x_1+b_2x_2+b_3x_3$	1.65%
$\hat{y}=a+b_1x_1+b_2x_2+b_3x_3+b_4x_1x_2+b_5x_1x_3+b_6x_2x_3+b_7x_1^2+b_8x_2^2+b_9x_3^2$	98.66%

(b) lineolic acid = 40, kerosene = 20, antioxidant = 5
$\hat{y} = 0.401+0.0011(40)-.0033(20)-.0046(5)=0.356$

The corresponding residual is:
$(y-\hat{y})=(.202-.356)=-0.154$

(c) lineolic acid = 40, kerosene = 20, antioxidant = 5

$\hat{y} = 0.1801$

The corresponding residual is:

$(y-\hat{y})=(.202-.1801)=0.0219$

(d) The larger residual magnitude based on $\hat{y} = a + b_1x_1 + b_2x_2 + b_3x_3$ is reasonable given the corresponding low coefficient of determination.

41. (a) Using Minitab:

Predictor	Coef	SE Coef	T	P
Constant	89.111	8.529	10.45	0.000
x1	-0.04962	0.01200	-4.14	0.001
x2	6.564	1.194	5.50	0.000
x3	-27.418	2.400	-11.43	0.000

S = 4.43208 R-Sq = 91.7% R-Sq(adj) = 90.2%

R^2 = 91.7% of the observed variation in y can be explained by the approximate relationship between y and the predictors x_1, x_2 and x_3.

(b) Using Minitab:

Term	Coef	SE Coef	T	P
Constant	55.7035	47.2484	1.17895	0.260
x1	0.0185	0.0846	0.21827	0.831
x2	8.7189	9.8318	0.88681	0.391
x3	-11.3128	21.5636	-0.52463	0.609
x1*x2	-0.0046	0.0167	-0.27663	0.786
x1*x3	-0.0330	0.0334	-0.98838	0.341
x2*x3	0.1050	3.3438	0.03140	0.975

S = 4.72891 R-Sq = 92.37% R-Sq(adj) = 88.84%

R^2 = 92.37% of the observed variation in y can be explained by the approximate relationship between y and the predictors x_1, x_2 and x_3 and the interactions.

(c) Using Minitab:

Term	Coef	SE Coef	T	P
Constant	81.2331	45.8093	1.77329	0.107
x1	0.1235	0.1116	1.10684	0.294
x2	-6.8372	10.2686	-0.66583	0.521
x3	-42.0354	19.4262	-2.16385	0.056
x1*x2	-0.0046	0.0124	-0.37388	0.716
x1*x3	-0.0330	0.0247	-1.33588	0.211
x2*x3	0.1050	2.4740	0.04244	0.967
x1*x1	-0.0001	0.0001	-1.13707	0.282
x2*x2	1.9445	0.9060	2.14635	0.057
x3*x3	10.2409	3.6944	2.77203	0.020

S = 3.49881 R-Sq = 96.79% R-Sq(adj) = 93.89%

R^2 = 96.79% of the observed variation in y can be explained by the approximate relationship between y and the predictors x_1, x_2 and x_3, the interactions, and quadratic predictors.

Section 3.6

43. (a) Exactly one car (x =1) and exactly one bus (y = 1) appear at the intersection .030 (i.e., 3%) of the time.

(b) At most one vehicle of each type ($x \le 1$ and $y \le 1$) appear at the intersection: p(0,0) + p(0,1) + p(1,0) + p(1,1) = .025 + .015 + .050 + .030 = .120 or, about 12% of the time.

(c) The number of cars equals the number of buses (x = y): p(0,0) + p(1,1) + p(2,2) = .025 + .030 + .050 = .105 or, about 10.5% of the time.

(d) Adding the rows yields the marginal distribution of the x values:

x :	0	1	2	3	4	5
p(x):	.05	.10	.25	.30	.20	.10

so, the mean number of cars is μ_x = 0(.05) + 1(.10) + 2(.25) + 3(.30) + 4(.20) + 5(.10) = 2.8.

(e) The number of vehicle spaces occupied is h(x,y) = x+3y. The mean number of spaces occupied is then, $\mu_{h(x,y)} = \sum\sum (x+3y)\cdot p(x,y)$ = [0 + 3(0)](.025) + [0+3(1)](.015) + …+ [5+3(2)](.020) = 4.90.

45. Summing across the rows (for x) and the columns (for y), the marginal distributions of x and y are:

x:	200	500		y:	0	250	500
$f_x(x)$:	.50	.50		$f_y(y)$:	.25	.25	.50

So, the means of the two distributions are: μ_x = 200(.50) + 500(.50) = 350 and μ_y = 0(.25) + 250(.25) + 500(.50) = 312.5. The covariance between the two variables is then: covariance(x,y) = $\sum\sum(x-\mu_x)(y-\mu_y)f(x,y)$, where f(x,y) is the joint mass function given in exercise 38. That is, covariance(x,y) = (200-350)(0-312.5)(.20) + (200-350)(250-312.5)(.10)+…+ (500-350)(500-312.4)(.30) = 9375.0.

Using the marginal distributions again, the standard deviations of each variable can be computed:

$\sigma_x^2 = \sum(x-\mu_x)^2 p(x)$ = $(200-350)^2(.50) + (500-350)^2(.50)$ = 22500, so σ_x = 150 and, $\sigma_y^2 = \sum(y-\mu_y)^2 p(y)$ = $(0-312.5)^2(.25) + (250-312.5)^2(.25) + (500-312.5)^2(.50)$ = 42,968.75, so σ_y = 207.289. The population correlation coefficient is then ρ = covariance/($\sigma_x\sigma_y$) = 9375/[(150)(207.289)] = .302. Because it is closer to 0 than to 1, this value of ρ does not suggest that there is a strong relationship between x and y.

Supplementary Exercises

47.

(a) b = .500 means that, on average, for each 1 °F increase in temperature there is about .500 unit increase in strength.

(b) The least squares line is $\hat{y}$ = -25.000 + .500x, so for x = 120, $\hat{y}$ = -25.000 + .500(120) = 35. The residuals for the values y = 40 and y = 29 are: y- $\hat{y}$ = 40-35 = 5 and y- $\hat{y}$ = 29-35 =-6. The residuals have different signs because one point (x=120, y=40) lies above the regression line and the other (x=120, y=29) lies below the line.

(c) Coefficient of determination = r^2 = 1 - (SSResid/SSTo) = 1 - (390.0/1060.0) = .632, or about 63.2%. This means that about 63.2% of the observed variation in strength can be attributed to the approximate linear relationship between strength and treatment temperature. This is a good value for r^2 (it is closer to 1 than to 0), but it might be possible to improve/increase r^2 by including some additional predictor variables and fitting a multiple regression model.

49. (a) Since stride rate is being predicted, y = stride rate and x = speed. Therefore, $SS_{xx} = \sum x_i^2 - (\sum x_i)^2/n$ = 3880.08 - $(205.4)^2/11$ = 44.7018, $SS_{yy} = \sum y_i^2 - (\sum y_i)^2/n$ = 112.681 - $(35.16)^2/11$ = .2969, and $SS_{xy} = \sum x_i y_i - (\sum x_i)(\sum y_i)/n$ = 660.130 - (205.4)(35.16)/11 = 3.5969. Therefore, b = SS_{xy}/SS_{xx} = 3.5969/44.7018 = .0805 and a = (35.16/11) - (.0805)(205.4/11) = 1.6932. The least squares line is then $\hat{y}$ = 1.6932 + .0805x.

(b) Predicting speed from stride rate means that y = speed and x = stride rate. Therefore,interchanging the x and y subscripts in the sums of squares computed in part (a), we now have SS_{xx} = .2969 and SS_{xy} = 3.5969 (note that SS_{xy} does not change when the roles of x and y are reversed). The new regression line has a slope of b = SS_{xy}/SS_{xx} = 3.5969/.2969 = 12.1149 and an intercept of a = (205.4/11) - (12.1149)(35.16/11) = -20.0514; that is, $\hat{y}$ = -20.0514 + 12.1149x.

(c) For the regression in part (a), r = 3.5969/[$\sqrt{44.7018}\sqrt{.2969}$] = .9873, so $r^2 = (.9873)^2 = .975$. For the regression in part (b), r is also equal to .9873 (since reversing x and y has no effect on the formula for r). So, both regressions have the same coefficient of determination. For the regression in part (a), we conclude that about 97.5% of the observed variation in rate can be attributed to the approximate linear relationship between speed and rate. In part (b), we conclude that about 97.5% of the variation in speed can be attributed to the approximate linear relationship between rate and speed.

51. (a) Substituting x = .005 into the least squares equation, a stress value of $\hat{y}$ = 88.791 + 5697.0(.005) - $328{,}161(.005)^2$ = 109.07 would be predicted.

(b) From the stress values (i.e, y values given in the exercise,) SSTo = $SS_{yy} = \sum y_i^2 - (\sum y_i)^2/n$ = 107,604 - $(1034)^2/10$ = 688.40. Subtracting predicted values from actual values gives the residuals: -3.16, -1.87, 5.07, 1.93, 2.84, -3.36, -1.48, -2.22, 2.07, and 0.20. The sum of squares of these residuals is SSResid = 73.711. The coefficient of determination r^2 = 1 - SSResid/SSTo = 73.711/688.40 = .893. Therefore, about 89.3% of the observed variation in stress can be attributed to the approximate linear relationship between stress sand strain.

(c) Substituting x = .03 into the least squares equation yields a predicted stress value of $\hat{y}$ = 88.791 + 5697.0(.03) - $328{,}161(.03)^2$ = -35.64, which is not at all realistic (since stress values can not be negative). The problem is that the value x = .03 lies far outside the region of x values that were used to fit the regression equation (the maximum x value used was x = .017). Extrapolation such as this is often unreliable (at best) and sometimes leads to ridiculous (i.e., impossible) predictions as in this case.

53. (a) The curvature that is apparent in the plot of y versus x (see below) indicates that merely fitting a straight-line to the data would not be the best strategy. Instead, one should search for some transformation of the x or y data (or both) that would give a more linear plot.

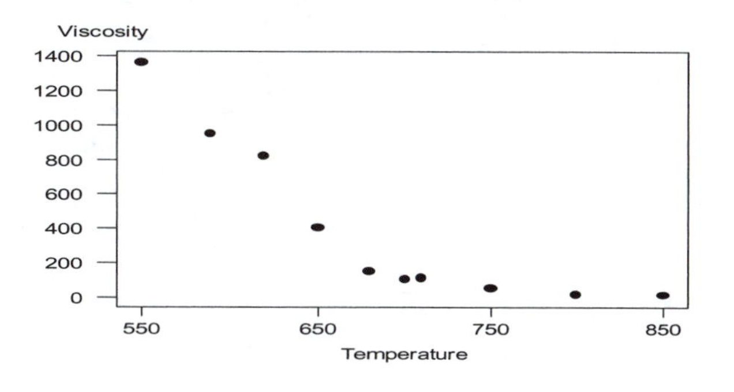

(b) The plot below shows the graph of ln(y) versus 1/x. Because it appears to be approximately linear, a straight-line fit to such data should provide a reasonable approximation to the relationship between the two variables.

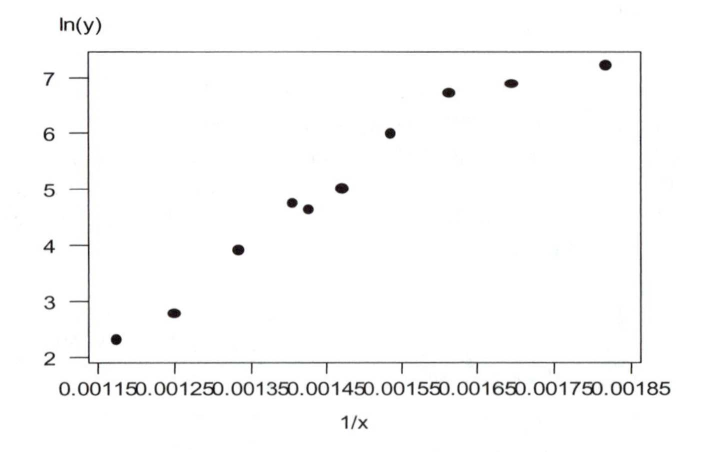

The following Minitab printout shows the results of fitting a regression line to the transformed data. From the printout, the prediction equation is $\ln(y) \approx -7.2557 + 8328.4(1/x)$. The r^2 value of 95.3% indicates that the fit is quite good. When temperature is 720 (i.e., $x = 720$), the equation gives a predicted value of $\ln(y) \approx -7.2557 + 8328.4(1/720) = 4.31152$. Exponentiating both sides gives a predicted y value of $y \approx e^{4.31152} = 74.6$.

```
The regression equation is
logey = - 7.26 + 8328 recipx

Predictor        Coef        StDev          T        P
Constant      -7.2557       0.9670      -7.50    0.000
recipx         8328.4        651.1      12.79    0.000

S = 0.3882        R-Sq = 95.3%      R-Sq(adj) = 94.8%

Analysis of Variance

Source        DF          SS          MS          F         P
Regression     1      24.663      24.663     163.63     0.000
Error          8       1.206       0.151
Total          9      25.869
```

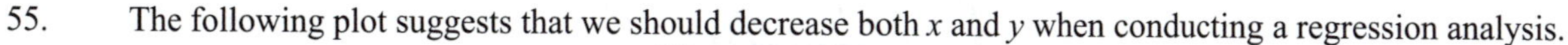

55. The following plot suggests that we should decrease both x and y when conducting a regression analysis.

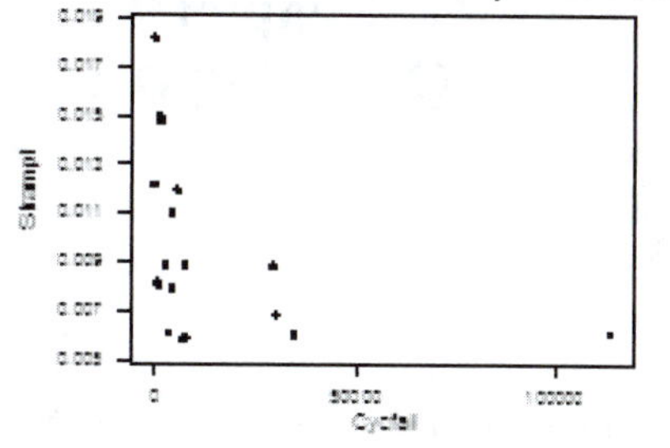

An appropriate curve regresses $\ln y$ on $\ln x$, and from MINITAB, we have $\hat{\ln}(y) = -3.74 - .124\ln(x)$. A measure of fit for this model is $r^2 = 46.9\%$. To predict strain amplitude when $x = 5000$, we first take $\ln 5000 = 8.517$, and so $\hat{\ln}(y) = -3.74 - .124\ln(5000) = -4.796$. So $\hat{y} = e^{-4.796} \approx .0083$.

```
The regression equation is
Log StrAmpl = - 3.74 - 0.124 Log Cycle

Predictor        Coef    SE Coef        T       P
Constant      -3.7372     0.2695   -13.87   0.000
Log Cycl     -0.12395    0.03199    -3.87   0.001

S = 0.2731      R-Sq = 46.9%      R-Sq(adj) = 43.8%

Analysis of Variance

Source          DF        SS        MS       F       P
Regression       1    1.1190    1.1190   15.01   0.001
Residual Error  17    1.2675    0.0746
Total           18    2.3865

Unusual Observations
Obs  Log Cycl  Log StrA       Fit    SE Fit  Residual   St Resid
  6       3.3   -4.0069   -4.1411    0.1698    0.1342     0.63 X

X denotes an observation whose X value gives it large influence.
```

Chapter 4
Obtaining Data

Section 4.1

1. Operational definitions are used to define measurement procedures. Benchmarks are objects or procedures used to compare two or more products or processes.

3. For example, you might use a definition such as: *Temperature at 2:00 pm in a fixed, unshaded area on top of City Hall.*

5. ISO ppm is an operational definition.

Section 4.2

7. There are several sampling methods that would work. Here is one of them.

   Divide the one-square-mile area of forest into 100 smaller regions each of equal size (each area would be $(1/10)^{th}$ of a mile by $(1/10)^{th}$ of a mile). Call each region a cluster. Then employ cluster sampling. Randomly sample n of these clusters. Within each sampled cluster study all of the trees that are growing.

9. (a) Both methods are capable of generating random samples from the block of trees.

   Research A's suggestion involves first thinking of the "population" as rows of trees. There are 40 rows and he recommends taking a random sample of 5 of the rows. Nest Researcher A must think of the "population" as the 25 trees within a row. He recommends taking a random sample of 6 of the 25 trees in each previously sampled rows. Using this method the chance a tree is selected is: $\left(\frac{5}{40}\right)\left(\frac{6}{25}\right)=.03$. Each tree has this same chance of selection, making this sampling scheme a random sampling scheme.

   Researcher B's suggestion is much simpler but also produces a random sample where the chance of selection is $\left(\frac{30}{1000}\right)=.03$ for each tree.

   (b) This type of sampling procedure is called stratified random sampling. Each quadrant is called a strata.

11. To generate a random integer between 1 and 1,000, type the following code in an Excel cell:

   ```
   = RANDBETWEEN (1,1000)
   ```

   Each time this command is used a new random integer will be generated. That is, sampling with replacement is used.

13. (a) $N = \sum_{i=1}^{k} N_i = \sum_{i=1}^{10} 1000 = 10(1000) = 10{,}000$

(b) Since we now require a confidence level of 90%, our z value is now $z = 1.645$. Without any previous information about π_i, we set $\pi_i = 0.5$ each stratum. Thus, $\sigma_i^2 = \pi_i(1-\pi_i) = .5(1-.5) = .25$ and $\sigma_i = \sqrt{.25} = .5$.

Under the assumption of equal unit costs and variances, our minimum sample size is:

$$n = \frac{\left(\sum_{i=1}^{k} N_i\sigma_i\right)^2}{N^2\left(\frac{B}{1.645}\right)^2 + \sum_{i=1}^{k} N_i\sigma_i^2} = \frac{\left(\sum_{i=1}^{10} (1000)(.5)\right)^2}{(10{,}000)^2\left(\frac{.03}{1.645}\right)^2 + \sum_{i=1}^{10}(1000)(.25)} = 699.12\text{, so round up to 700.}$$

Now, each stratum sample size is $n_i = n(N_i / N)$, but $N_i / N = 1000/10{,}000 = .10$ for each stratum. So $n_i = (700)(.10) = 70$ for each stratum.

(c) $p_1 = 5/70$, $p_2 = 4/70$, ..., $p_{10} = 8/70$

$$p_{\text{str}} = \sum_{i=1}^{10} p_i\left(\frac{N_i}{N}\right) = \sum_{i=1}^{10} p_i(.10) = .10\sum_{i=1}^{10} p_i = .10\left(\frac{5}{70} + \frac{4}{70} + \cdots + \frac{8}{70}\right) = .10\left(\frac{55}{70}\right) \approx .07857$$

(d) $$s_p^2 = \frac{1}{N^2}\sum_{i=1}^{k} N_i^2\left(\frac{N_i - n_i}{N_i}\right)\left(\frac{p_i(1-p_i)}{n_i - 1}\right) = \underbrace{\frac{1}{N^2}N_i^2\left(\frac{N_i - n_i}{N_i}\right)\left(\frac{1}{n_i - 1}\right)}_{\text{since each stratum has the same } N_i \text{ and } n_i}\sum_{i=1}^{10} p_i(1-p_i)$$

$$= \frac{N_i}{N^2}\left(\frac{N_i - n_i}{n_i - 1}\right)\sum_{i=1}^{10} p_i(1-p_i) \approx \frac{1000}{10{,}000^2}\left(\frac{1000-70}{70-1}\right)(.7153061224) = 9.64108\times 10^{-5}$$

So $s_p = \sqrt{9.64108\times 10^{-5}} \approx .0098189$

15. (a) Let w_i denote the weight for stratum i. Since all c_i are equal, we can multiply the numerator and denominator by $\sqrt{c_i}$ to simplify:

$$w_i = \left(\frac{N_i\sigma_i/\sqrt{c_i}}{N_1\sigma_1/\sqrt{c_1} + N_2\sigma_2/\sqrt{c_2} + \cdots + N_k\sigma_k/\sqrt{c_k}}\right)\frac{\sqrt{c_i}}{\sqrt{c_i}}$$

$$= \frac{N_i\sigma_i}{N_1\sigma_1 + N_2\sigma_2 + \cdots + N_k\sigma_k} = \frac{N_i\sigma_i}{\sum_{i=1}^{k} N_i\sigma_i}.$$ This is exactly the Neyman allocation.

(b) Since all unit sampling costs are equal, then all c_i are equal for $i = 1, \ldots, k$. Moreover, since all strata variances are equal, then each of the σ_i's are equal. Some algebra work gives:

$$w_j = \left(\frac{N_j \sigma_j / \sqrt{c_j}}{N_1 \sigma_1 / \sqrt{c_1} + N_2 \sigma_2 / \sqrt{c_2} + \cdots + N_k \sigma_k / \sqrt{c_k}} \right) \left(\frac{\sqrt{c_i} / \sigma_i}{\sqrt{c_i} / \sigma_i} \right) = \frac{N_j}{N_1 + N_2 + \cdots + N_k} = \frac{N_j}{N}.$$ Now let the total sample size be n. Since the appropriate weight is $w_j = N_j / N$, the appropriate sample size for stratum j is $n_j = n \cdot w_j = n(N_j / N)$, which is precisely what is specified by "proportional" allocation.

17. In general, we note that the total sample size for a given confidence level is $n = \frac{\sum_{i=1}^{k} N_i^2 \sigma_i^2 / w_i}{N^2 \left(\frac{B}{z \text{ critical}} \right)^2 + \sum_{i=1}^{k} N_i \sigma_i^2}$, where z critical denotes the appropriate z value for the given confidence level. We know that as we increase the confidence level, we necessarily increase the z value. In particular, increasing the confidence from 95% to 99% results in increasing the critical z value from 1.96 to 2.575. Now since the z value is in the denominator of the denominator, we ultimately increase the magnitude of the *numerator* as we increase the z value. Hence, the required sample size increases in order to achieve higher confidence. This intuitively makes sense.

Section 4.3

19. The purpose of replicated measurements is two fold:
    1) biases tend to be eliminated when several measurements are averaged; but more importantly,
    2) the variation between repeated measurements gives a measure of experimental error.

21. (a) Variation in fuel efficiency between 100-mile segments can be quantified if one measures fuel efficiency every segment. That is, if the tool of replication is used. If one measures efficiency at the end of the 500-mile course, there is no measure of experimental error.

    (b) The researcher should consider specifying those variables that may effect fuel efficiency. Examples include: type and condition of the vehicle, tire pressure, driving speed and style, environmental conditions, etc…

    (c) To draw conclusions about the effectiveness of the new fuel additive, the researcher may want to assess the effectiveness under different experimental conditions by introducing experimental factors and blocking variables. For example, the researcher may wish to determine the effect that "vehicle type" has on the response. The factor "vehicle type" might include the levels: compact, mid-sized, full-sized, sport-utility, etc… By investigating the "vehicle type" variable, the researcher can make generalization about different vehicle types.

    Another example could address the creation of a blocking variable. Suppose the researcher proposed that a different driver drive the vehicle each 100 miles. By including this "person" blocking variable, the researcher can make generalizations about drivers in general.

23. Two basic experimental design principles are violated; replication and randomization. First, even though the set-up procedures may be lengthy and tedious, they should be performed each time an experimental run is called for. Otherwise, true replication cannot be achieved.

Second, there may be an effect from one day to the next and this possible "day effect" is confounded with the fact that a different lab assistant conducted the experiments on each day, making it impossible for us to separate the possible "day effect" from the possible "person effect". Randomization would require that the lab assistants be randomly selected for each of the 12 experimental runs.

Section 4.4

25. An estimate of the accuracy requires that the mean be computed:

$\bar{x} = .3024$

Recall that the true value, x, is .300. So, accuracy $= (\bar{x} - x) = (.3024 - .300) = .0024$

An estimate of the precision requires that the standard deviation be computed:

Precision = s = .0024083

27. (a)

Measurement, m	Relative Error
.301	.333%
.303	1%
.299	-.333%
.305	1.67%
.304	1.33%

(b) The maximum absolute error you would expect in a measured reading of 70 degrees Fahrenheit from this thermometer is:
(70)(.04) = 2.8 degree Fahrenheit

29. (a) Here's a Youden plot of the data:

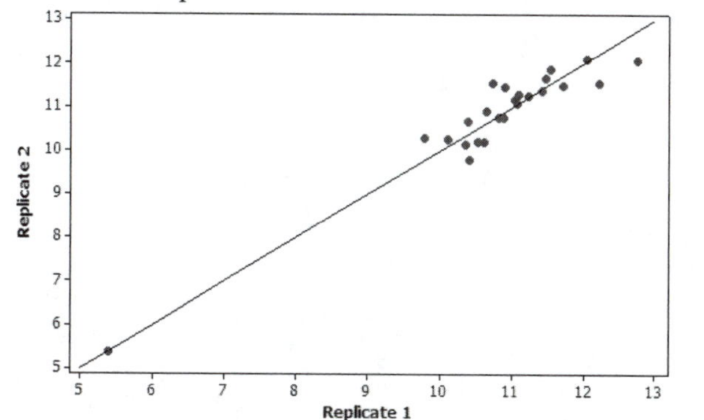

(b) The Youden plot for this data shows many points near the 45 degree line, indicating that several of the laboratories are following slightly different versions of the test procedure. Lab 19 in the bottom left corner is using procedures that are systematically different from the other labs; most of the other labs are following slightly different versions of measuring MFI.

Supplementary Exercises

31. Suppose you take a random sample of size n with replacement. Then according to Rule 1 in Section 4.2, the complement of this random sample is also a random sample. Notice that the complement will contain no duplicates. Finally, using Rule 1 again, the complement of the complement will be a random sample and is equivalent to the original random sample but with duplicates discarded (i.e., a random sample without replacement).

33. (a) The background samples of air would be used as a benchmark of the ambient levels of Cr(VI) in the air. Once air samples at chromite ore plants are taken, the background samples can be compared to the plant samples in order to estimate the increase in Cr(VI) pollutant at chromite ore plants.

    (b) ASTM Standard Test Method D5281-92 is the operational definition for how measurements are to be made. By using this standard method to measure Cr(VI) concentration, the authors hope to reduce measurement variation so that any changes in Cr(VI) concentrations can be attributed to the chromite ore plants and not to variation in the measurement system.

    (c) The location at which an air sample is taken can be considered an experimental factor (i.e., independent variable).

    The six sampling periods illustrate the experimental principle of replication.

    Distinguishing between wet and dry days constitutes blocking.

Chapter 5
Probability and Sampling Distributions

Section 5.1

1. (a) Sampling without replacement means that no repeated items will occur in any sample. There are 10 possible such samples of size 3:

   {a,b,c}, {a,b,d}, {a,b,e}, {a,c,d}, {a,c,e}, {a,d,e}, {b,c,d}, {b,c,e}, {b,d,e}, {c,d,e}.

   As you learned in the discussion of the binomial distribution, the number of ways to choose a sample of 3 distinct items from a list of 5 items is given by $\binom{5}{3} = 10$, which shows that the list above is indeed complete.

   (b) A contains 3 samples from the list in (a); i.e., A = { {a,b,c}, {a,c,d}, {a,c,e}}

   (c) The complement of A is A′ = {{a,b,d}, {a,b,e}, { {a,d,e}, {b,c,d}, {b,c,e}, {b,d,e}, {c,d,e}}

3. To envision the events A and B, it helps to draw a number line with the integers representing the possible numbers of defective items:

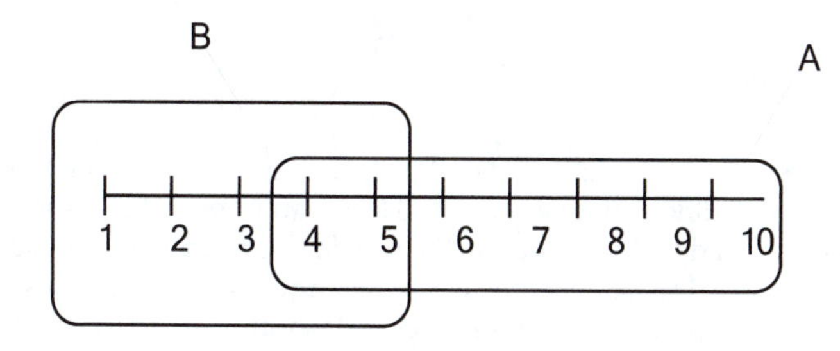

   (a) The event *A and B* consists of the integers {4, 5}. That is, *A and B* is the event 'either 4 or 5 defectives in the sample'.

   (b) The event *A or B* consists of all ten integers. There are many ways to describe this event. One description of *A or B* is the event 'there is at least one defective in the sample'.

   (c) The complement of A consists of the integers {1, 2, 3}. In words, A′ is the event 'there are at most 3 defectives in the sample'.

5. The tree diagram is shown below. Note that it is not necessary for all the branches to be of the same length. That is, some branches may stop early in the tree, while others may extend through several additional branching points.

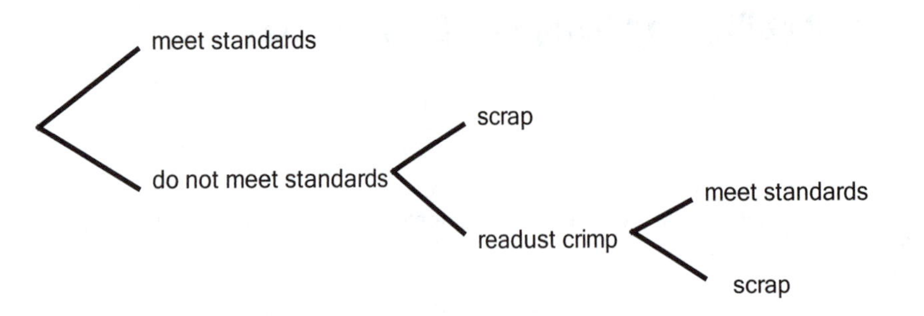

7. The event *A and B* is the shaded area where A and B overlap in the following Venn diagram. Its complement consists of all events that are either not in A or not in B (or not in both). That is, the complement can be expressed as *A′ or B′.*

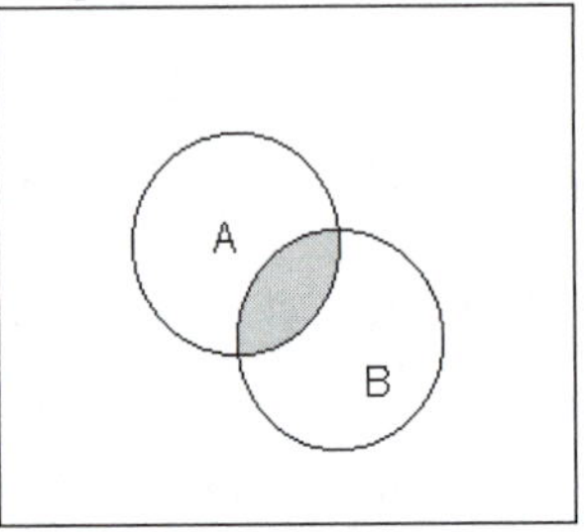

Section 5.2

9. (a) The total of 724 +751 solder joints overstates the actual number found, since 316 solder joints were found by *both* inspectors. To avoid such double-counting (i.e., the 316 is part of both the 724 joints found by inspector A and the 751 joints found by Inspector B), 316 should be subtracted from the raw totals, which means that 724 + 751 - 316 = 1,159 *distinct* joints were identified by the inspectors together. The important point to note in this problem is that the events 'Inspector A finds a defective solder joint' and 'Inspector B finds a defective solder joint' are *not* necessarily mutually exclusive, so we can not simply add the numbers of joints (or, equivalently, the probabilities of finding defective joints) for both inspectors.

   (b) *A and B′* contains 724 - 316 = 408 solder joints.

11. Letting A_i denote the event that the ith component fails (note that this is different from the definition of A_i used in problem 5.10), the probability that the entire series system fails is denoted by $P(A_1$ or A_2 or A_3 or ... or $A_k)$. Given that each $P(A_i) = .01$, the problem states that $P(A_1$ or A_2 or A_3 or ... or $A_k) \leq P(A_1) + P(A_2) + P(A_3) + \ldots + P(A_k) = 5(.01) = .05$. That is, there is *at most* a 5% chance of system failure.

Section 5.3

13. (a) Note that this question is equivalent to asking 'what is the probability A (or B, etc.) is chosen *given* that we know E is not chosen'. That is, the new probabilities are now conditional probabilities, where the conditioning event is that E is not chosen. Therefore,

$P(A \mid E') = P(A \text{ and } E')/P(E') = P(A)/P(E') = .20/(1-.10) = .20/.90 = 20/90.$
$P(B \mid E') = P(B \text{ and } E')/P(E') = P(B)/P(E') = .25/(1-.10) = .25/.90 = 25/90.$
$P(C \mid E') = P(C \text{ and } E')/P(E') = P(C)/P(E') = .15/(1-.10) = .15/.90 = 15/90.$
$P(D \mid E') = P(D \text{ and } E')/P(E') = P(D)/P(E') = .30/(1-.10) = .30/.90 = 30/90.$

Note that the four probabilities above do add exactly to 1, reflecting the fact that A, B, C, and D are the only four choices possible now that company E has been eliminated.

(b) The probability of *not* choosing companies B, D, or E is 1 -(.25+.30+.10). Following the same reasoning as in part (a), the two revised probabilities of choosing A or C are:

$P(A \mid B, D, E \text{ not chosen}) = P(A)/(1-(.25+.30+.10)) = .20/.35 = 20/35.$
$P(C \mid B, D, E \text{ not chosen}) = P(C)/(1-(.25+.30+.10)) = .15/.35 = 15/35.$

Again, the revised probability sum to 1 since there are now only two choices allowed.

15. It is helpful to first put the probabilities in this exercise on the tree diagram.

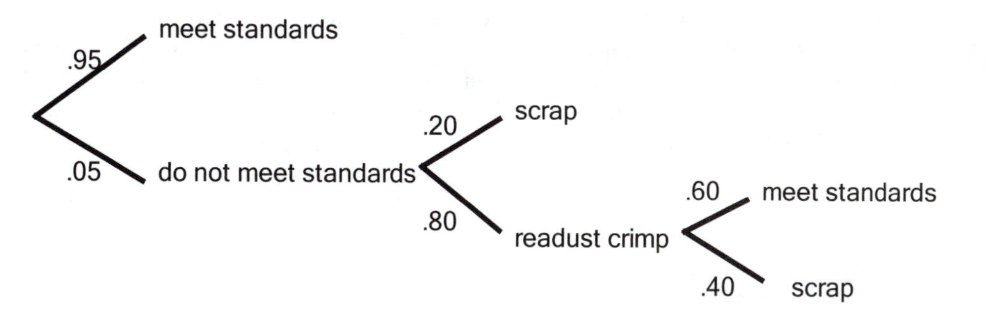

(a) The branches associated with passing the second inspection have probabilities .80 (those that require recrimping) and .60 (the recrimped fasteners that pass inspection). Recall that probabilities of following any branch on a tree diagram is simply the product of the probabilities of the sub-branches. Therefore, (.80)(.60) = .48, or, 48% of the fasteners that fail the initial inspection will go on to pass the second inspection. Be careful, this problem asks for the percentage of *failed* fasteners that eventually pass inspection, not the percentage of *all* fasteners initially submitted for inspection.

(b) From the tree diagram, the fasteners that pass inspection are those that either pass on first inspection (95%), or pass on second inspection after recrimping. These two events are mutually exclusive, so their probabilities can be added. Following the branches through the tree, the proportion of *all* fasteners that pass the second inspection is (.05)(.80)(.60) = .024. Therefore, 95% + 2.4% = 97.4% of all fasteners eventually pass inspection.

(c) Let I denote the event that a fastener passes inspection and let F denote the event that a fastener passes the first inspection. Then the problem asks for the conditional probability $P(F \mid I)$. Using the conditional probability formula, $P(F \mid I) = P(F \text{ and } I)/P(I)$. Note that P(I) = .024 was calculated in part (b) of this question. Also, the event *F and I* can be simplified to F; i.e., *F and I* $= F$, so P(*F and I*) = P(F) = .95. The desired probability is then $P(F \mid I) = .95/.974 = .9754$, or 97.54%.

17. The probabilities of independent events A and B must satisfy the equation P(*A and B*) = P(A)·P(B). If A and B were also mutually exclusive, then P(*A and B*) would equal 0, which would mean that P(A)·P(B) = P(*A and B*) = 0. This would require that at least one of A or B have zero probability of occurring. Although this is technically possible, most events of interest have non-zero probabilities, making P(A)·P(B) non-zero. It is therefore impossible for independent events with non-zero probabilities to be mutually exclusive.

19. (a) Since people's blood type is independent, $P(1_A \text{ and } 2_A) = P(1_A)P(2_A) = (.42)(.42) = .1764$

(b) $P(1_B \text{ and } 2_B) = (.10)^2 = .01$
$P(1_{AB} \text{ and } 2_{AB}) = (.04)^2 = .0016$
$P(1_0 \text{ and } 2_0) = (.44)^2 = .1936$

(c) P(matching blood types) = P(both have A) + P(both have B) + P(both have AB) + P(both have 0) = .1764 + .01 +.0016 + .1936 = .3816

(d) Discriminating Power = P(the 2 people do not have matching blood types)
= 1 – P(the 2 people do have matching blood types)
= 1 – (.3816) = .6184

21. Let A denote the event that components 3 and 4 *both* work correctly and let B denote the event that *at least one* of components 1 or 2 works correctly. Then P(systems works) = P(A or B). From the general addition law, P(A or B) = P(A) +P(B) - P(A and B). Because all components act independently of one another, P(A) = P(3 and 4 work) = P(3 works)·P(4 works) = (.9)(.9) = .81. P(B) = P(1 or 2 works) = P(1 works) + P(2 works) - P(1 and 2 work) = .9 + .9 - (.9)(.9) = .99. Finally, events A and B are independent since A involves only components 3 and 4, whose actions are independent of components 1 and 2, so P(A and B) = P(A)·P(B) = (.81)(.99) = .8019. Therefore, P(A or B) = P(A) +P(B) - P(A and B) = .81 + .99 - .8019 = .9981.

23. Letting D denote the event that a person has the disease and letting T be the event that the test indicates a person has the disease (usually called a 'positive' test result), a tree diagram summarizes the information in the problem:

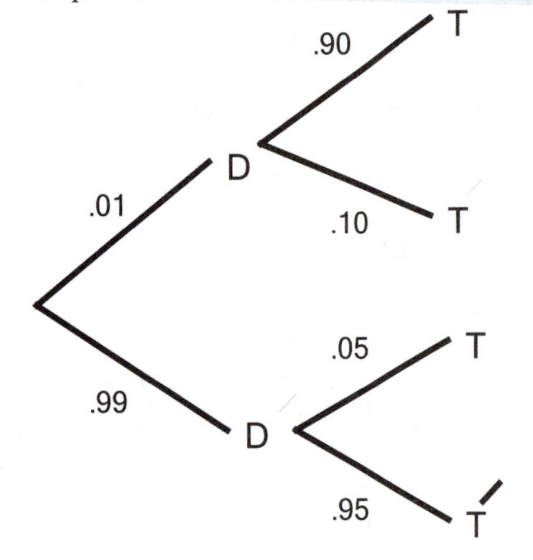

(a) For people *with* the disease, the probability of two positive test results is $(.90)^2 = .81$; similarly, for those with the disease, the probability of two negative test results is $(.10)^2 = .01$. The probability of two similar results is then .81+.01 = .82. When applied to people who do not have the disease, the same argument shows that the probability of two similar test results is $(.05)^2 + (.95)^2 = .905$. Thus, 91% of the 1% of people who have the disease will get two identical test results, while 90.5% of the 99% disease-free people will get two identical test results. The proportion of the entire population that gets two identical test results is then (.82)(.01) + (.905)(.99) = .90415.

(b) P(both tests positive) = P(both positive and D *or* both positive and D′) = P(both positive and D) + P(both positive and D′) = P(both positive | D)·P(D) + P(both positive | D′)·P(D′) = $(.90)^2(.01) + (.05)^2(.99)$ = .0081 + .0022475 = .010575. Therefore, P(D | both tests positive) = [P(both positive | D)·P(D)]/P(both tests positive) = .0081/.010575 = .766, or, about 76.6%.

25. The event *A′ and B* is shaded in the following Venn diagram:

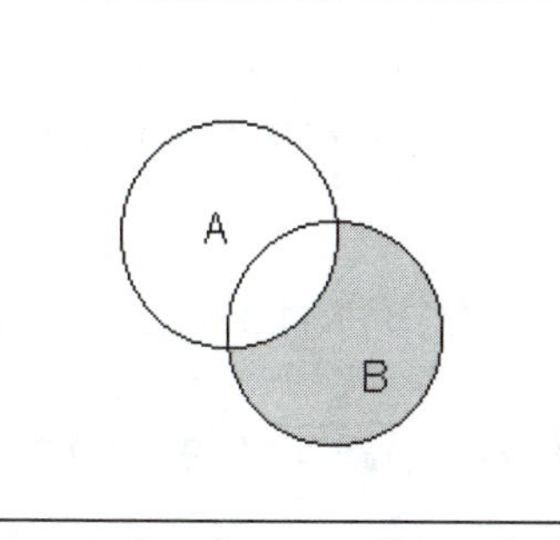

From the diagram you can also see that the mutually exclusive events *A and B* and *A′ and B* comprise event B; i.e., B = (*A and B*) or (*A′ and B*). Using the addition law for exclusive events, P(B) = P(*A and B*) + P(*A′ and B*), which can be rearranged as P(*A′ and B*) = P(B) - P(*A and B*). Using the fact that A and B are independent, P(*A and B*) = P(A)· P(B), so, P(*A′ and B*) = P(B) - P(*A and B*) = P(B) - P(A)· P(B) = [1- P(A)]·P(B) = P(A′)·P(B), which shows that A′ and B are independent.

Section 5.4

27. (a) Discrete; because x has a finite, countable number of values.
(b) Continuous; because concentrations can have any conceivable value in an interval.
(c) Discrete; the total number of bolts is finite, which means that the possible number of oversized bolts as well as the proportion of oversized bolts can only have a finite number of possible values.
(d) Discrete; the number of errors must be finite.
(e) Continuous; because strength measurements can conceivably have any value in an interval of real numbers.
(f) Continuous; because time can take any value in an interval of real numbers.
(g) Discrete; because the number of customers must be finite.

29. (a) x is a discrete random variable, so

$$\mu = \sum_x xp(x) = (1)(.2) + (2)(.4) + (3)(.3) + (4)(.1) = 2.3.$$

(b) $\sigma^2 = \sum_x (x-\mu)^2 p(x)$ = $(1\text{-}2.3)^2(.2) + (2\text{-}2.3)^2(.4) + (3\text{-}2.3)^2(.3) + (4\text{-}2.3)^2(.1)$
= .81.

(c) The expected number of pounds shipped should be 5μ; i.e., the product of 5 lbs/order times the expected number of orders shipped, μ. Therefore, the expected number of pounds remaining after an order is shipped is 100 - 5μ = 100 - 5(2.3) = 88.5 lbs.

31. (a) y is a discrete random variable, so $\sum_y p(y)$ must equal 1. Using the formula for p(y), we have k(1) + k(2) + k(3) + k(4) + k(5) = 1, or, 15k = 1, so k = 1/15.

(b) P(at most three forms required) = P(y ≤ 3) = p(1) + p(2) + p(3) = 1(1/15) +2(1/15) + 3(1/15) = 6/15 = 2/5 = .40, or 40%.

(c) $\mu = \sum_y yp(y)$ = 1(1/15) + 2(2/15) + 3(3/15) +4(4/15) + 5(5/15) = 55/15 = 11/3.

(d) $\sigma^2 = \sum_y (y-\mu)^2 p(y)$ = $(1\text{-}11/3)^2(1/15) + (2\text{-}11/3)^2(2/15) + (3\text{-}11/3)^2(3/15) + (4\text{-}11/3)^2(4/15) + (5\text{-}11/3)^2(5/15)$ = 14/9 = 1.5555. Therefore, $\sigma = \sqrt{14/9}$ = 1.2472.

33. Let x = the number of defective solder joints. n = 285 and $\pi = .01$

(a) Recall from Chapter 2 that: $\mu = n\pi$ for a binomial random variable.
So, $\mu = (285)(.01) = 2.85$
Recall also that: $\sigma = \sqrt{n\pi(1-\pi)}$
So, $\sigma = \sqrt{(285)(.01)(.99)} = 1.6797$

(b) For a PCB to be defect-free, all 285 solder joints must be defect-free.

So: $\begin{pmatrix} \textit{proportion of all} \\ \textit{PCBs that are defect free} \end{pmatrix} = P(x=0) = (.99)^{285} = .05702$

(c) $P(x \geq 2) = 1 - P(x < 2) = 1 - P(x \leq 1) = 1 - [P(x=0) + P(x=1)]$

$$= 1 - \left[(.99)^{285} + \binom{285}{1} (.01)^1 (.99)^{284} \right] = 1 - [.05702 + .16415] = .77883$$

35. (a) Let x = 'number of breakthroughs'. Then x has a binomial mass function with n = 100 and π = .50, so μ = np = 100(.50) = 50.

(b) Using the normal approximation (with continuity correction) to binomial along with the fact that $\sigma = \sqrt{n\pi(1-\pi)} = 5$ gives $P(x \geq 60) \approx P(z \geq \frac{59.5-50}{5}) = P(z \geq 1.90) = 1 - P(z \leq 1.90) = 1 - .9713 = .0287$.

37. (a) x = 'number of defectives in a sample' has a binomial mass function with n = 10 and $\pi = .10$. Therefore, using Appendix Table II, P(accepting a lot) = $P(x \leq c) = P(x \leq 1) = .349 + .387 = .736$, so the probability of rejecting the lot is $1 - .736 = .264$.

(b) For a shipment with no defectives, $\pi = 0$, so the number of defectives in <u>any</u> sample must be x = 0, so it is a certainty that $x \leq 1$; that is, P(accepting such a lot) = $P(x \leq 1) = 1$.

(c)

π	**P(x ≤ 1) = P(x=0) + P(x =1)**	**P(reject lot) = 1 - P(x ≤ 1)**
.05	.914 = .599 + .315	.086 = 1 - .914
.20	.375 = .107 + .268	.625 = 1 - .375
.50	.011 = .001 + .010	.989 = 1 - .011

(d) The following Minitab plot shows the OC curve for a sampling plan with n =10, c = 1:

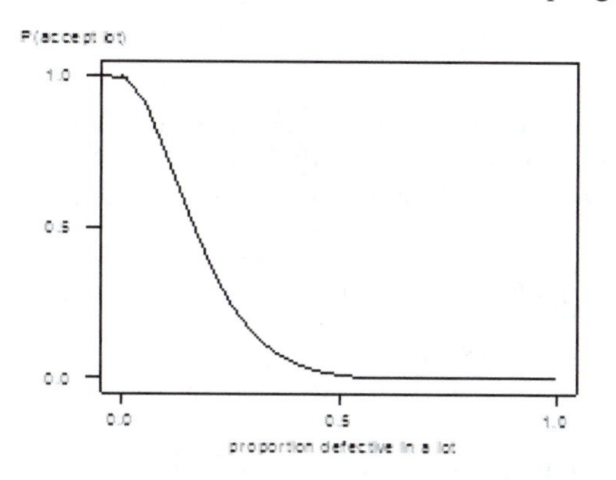

39. (a) x = 'number of correct answers' has a binomial mass function with n = 25 and $\pi = 1/5 = .20$.

(b) $\mu = 25(1/5) = 5$ and $\sigma = \sqrt{n\pi(1-\pi)} = 2$.

(c) Using Table II, the closest integer score S that satisfies $P(x \geq S) = .01$ is S = 11 for which $P(x \geq 11) = .004 + .002 = .006$.

41. (a) x = 'lifetime of an electronic component' has an exponential density with density function $f(x) = \lambda e^{-\lambda x}$ (for $x \geq 0$). In this problem $\lambda = 1/\mu = 1/\text{MTBF} = 1/500$. Therefore, the median $\tilde{\mu}$ of this random variable satisfies the equation:

$$.5 = \int_0^{\tilde{\mu}} \lambda e^{-\lambda x} dx = \left[-e^{-\lambda x}\right]_0^{\tilde{\mu}} = -[1 - e^{-\lambda\tilde{\mu}}].$$ That is, $-.5 = 1 - e^{-\lambda\tilde{\mu}}$, so $e^{-\lambda\tilde{\mu}} = 1 - .5 = .5$. Taking natural logarithms of both sides gives $-\lambda\tilde{\mu} = \ln(.5)$, or, $\tilde{\mu} = -\ln(.5)/\lambda = .69315(500) = 346.57$ hours.

(b) From the answer to part (a), $\tilde{\mu}$ = .69315/λ = .69315μ, which is less than μ.

(c) $\tilde{\mu}$ = .69315μ, or approximately, $\tilde{\mu}$ = .693MTBF.

43. (a) P(x = 5) = .20 + .15 + .05 = .40
P(x = 6) = .10 + .15+ .10 = .35
P(x = 7) = .10 + .10 + .05 = .25

(b) P(y = 10) = .20 + .15 + .05 = .40
P(y = 15) = .15 + .15 + .10 = .40
P(y = 20) = .05 + .10 + .05 = .20

(c) The (x,y) pairs that satisfy x + y ≤ 21 are: (5,10), (5,15), (6,10), (6,15), and (7,10), so P(x + y ≤ 21) = .20 + .15 + .10 + .15 + .10 = .95.

(d) x and y are not independent because P(x = 5 and y = 10) = .20 , whereas P(x = 5)P(y = 10) = (.40)(.40) = .16, so P(x=5 and y=10) ≠ P(x=5)P(x=10).

Section 5.5

45. (a) x has a binomial distribution with n = 5 and π = .05. Writing P(.05-.01 ≤ p ≤ .05+.01) in terms of x, we find P(.05-.01 ≤ x/5 ≤ .05+.01) = P((.04)5 ≤ x ≤ (.06)5) = P(.2 ≤ x ≤ .3). Because x can only have integer values, there is no x between .2 and .3, so the probability of this event is 0.

(b) For n= 25, P(.04 ≤ p ≤ .06) = P((.04)25 ≤ x ≤ (.06)25) = P(1 ≤ x ≤ 1.5) =
$P(x = 1) = \binom{25}{1}(.05)^1(.95)^{24} = 0.36498.$

(c) For n= 1005, P(.04 ≤ p ≤ .06) = P((.04)100 ≤ x ≤ (.06)100) = P(4 ≤ x ≤ 6). Using the normal approximation to the binomial, with μ = nπ = 100(.05) = 5 and σ^2 = nπ(1-π) = 100(.05)(.95) = 4.75 and σ = 2.7195: $P(4 \le x \le 6) \approx P(\frac{4-5}{2.7195} \le z \le \frac{6-5}{2.1795}) = P(-.46 \le z \le .46) = .3544$. Using the *continuity correction* makes a substantial difference in this problem because the interval from 4 to 6 contains the mean of the distribution (and hence, a large amount of the probability): $P(4 \le x \le 6) \approx P(\frac{3.5-5}{2.7195} \le z \le \frac{6.5-5}{2.1795}) = P(-.69 \le z \le .69) = .5098$.

47. (a) x = 'disconnect force' has a uniform distribution on the interval [2,4]. M is the maximum of a sample of size n = 2 from the uniform density on [2,4]. The larger of two items randomly selected from the interval [2,4] should, on average, tend to be closer to the upper end of the interval.

(b) Using the same reasoning as in part (a), the largest value in a sample of n = 100 will, most likely, be even closer to the upper end of the interval [2,4] than is the largest value in a sample of size n =2. So the average of all M's based on n=100 ought to be larger than the average value of all M's based on n = 2.

(c) For larger samples (e.g., n = 100), the maximum values will usually be fairly close to the upper endpoint of 4, which means that the variability amongst such values will tend to be small. For smaller samples (e.g., n = 2), it is easier for the value of M to wander over the interval [2,4], which means that the variability among these values will be larger than for n = 100.

Section 5.6

49. (a) $\mu_p = \pi = .80$. $\sigma_p = \sqrt{\frac{\pi(1-\pi)}{n}} = \sqrt{\frac{.80(1-.80)}{25}} = .08$

(b) Since 20% do not favor the proposed changes (so $\pi = .20$), the mean & standard deviation of the sampling distribution of this proportion are $\mu_p = \pi = .20$ and $\sigma_p = \sqrt{\frac{\pi(1-\pi)}{n}} = \sqrt{\frac{.20(1-.20)}{25}} = .08$.

(c) For n = 1000 and $\pi = .80$, $\mu_p = \pi = .80$. $\sigma_p = \sqrt{\frac{\pi(1-\pi)}{n}} = \sqrt{\frac{.80(1-.80)}{100}} = .04$. Notice that it was necessary to quadruple the sample size (from n=25 to n=100) in order to cut σ_p in half (from $\sigma_p = .08$ to $\sigma_p = .04$).

51. (a) n = 16, $\mu = 12$ and $\sigma = .04$, so $P(11.99 \le \bar{x} \le 12.01) =$

$$P\left(\frac{11.99-12}{.04/\sqrt{16}} \le z \le \frac{12.01-12}{.04/\sqrt{16}}\right) =$$

$P(-1 \le z \le 1) = .8413 - .1587 = .6826$.

(b) n = 64, $\mu = 12$ and $\sigma = .04$, so $P(11.99 \le \bar{x} \le 12.01) =$

$$P\left(\frac{11.99-12}{.04/\sqrt{64}} \le z \le \frac{12.01-12}{.04/\sqrt{64}}\right) =$$

$P(-2 \le z \le 2) = .9772 - .0228 = .9544$.

53. (a) x = 'lifetime of battery" has a normal density with $\mu = 8$ hours and $\sigma = 1$ hour. Therefore, P(average of 4 exceeds 9 hours) = $P(\bar{x} > 9) = P\left(z > \frac{9-8}{1/\sqrt{4}}\right) =$

$P(z > 2) = 1 - P(z<2) = 1 - .9772 = .0228$.

(b) Having the total lifetime of 4 batteries exceeds 36 hours is the same thing as having their average exceed 9, so the probability of this event is .0228, the same as in part (a).

(c) $.95 = P(T > T_0) = P(T/4 > T_0/4) = P(\bar{x} > T_0/4) = P\left(z > \frac{T_0/4-8}{1/\sqrt{4}}\right) =$

$P(z > T_0/2-16)$. For a standard normal distribution, $P(z > 1.645) \approx .95$, so we must have $T_0/2-16 = 1.645$, which gives $T_0 = 8 + 1.645/2 = 8.8225$ hours.

(d) Let Y = "replacement cost". Then Y is a discrete random variable with values \$0 and \$3 and corresponding probabilities .95 and .05. The expected value of Y (i.e., the average replacement cost per package) is μ = \$0(.95) + \$3(.05) = \$0.15 per package.

55. (a) x = 'sediment density' has a normal distribution with $\mu = 2.65$ and $\sigma = .85$), so $P(\bar{x} \le 3.00) = P(z \le \frac{3.00-2.65}{.85/\sqrt{25}}) = P(z \le 2.06) = .9803$. Similarly, $P(2.65 \le \bar{x} \le 3.00) = P(0 \le z \le 2.06) = .9803 - .5000 = .4803$.

(b) For any n, $P(\bar{x} \le 3) = P(z \le \frac{3.00-2.65}{.85/\sqrt{n}}) = P(z \le .41176\sqrt{n})$. Since the value of $z \approx 2.326$ is associated with a cumulative probability of .99, we equate $.41176\sqrt{n}$ and 2.326 and find $n = (2.326/.41176)^2 = 31.91$, or, about n = 32.

57. (a) Let p = 'proportion of resistors exceeding 105 Ω'. Then the sampling distribution of p is approximately normal with $\mu_p = \pi = 0.02$ and $\sigma_p = \sqrt{\frac{\pi(1-\pi)}{n}} = \sqrt{\frac{.02(1-.02)}{100}} = 0.014$.

(b) $P(p < .03) = P(z < \frac{.03-.02}{.014}) = P(z < .71) = 0.7611$.

59. (a) Let x = 'particle radius'. Then y = ln(x) has a normal distribution with $\mu = -2.62$ and $\sigma = .788$, so $\mu_x = \exp(\mu + \sigma^2/2) = \exp(-2.62 + (.788)^2/2) = \exp(-2.3095) = 0.099308$.

(b) $P(x > .12) = P(\ln(x) > \ln(.12)) = P(y > -2.1203) = P(z > \frac{-2.1203-(-2.62)}{.788}) = P(z > .63) = 1 - P(z \le .63) = 1 - .7357 = 0.2643$.

Supplementary Problems

61. (a) The sum of the parcel areas is 15+20+...+20 = 90, so $P(B_1 \text{ or } B_2 \text{ or } B_3) = P(B_1) + P(B_2) + P(B_3) = 15/90 + 20/90 + 25/90 = 60/90 = 2/3$.

(b) $P(B_5') = 1 - P(B_5) = 1 - 20/90 = 7/9$.

63. Let B_i denote the event that the ith battery operates correctly. Therefore, $P(B_i) = .95$ for each i = 1,2,3,4. Then, P(tool fails) = P(at least one battery fails) = 1 - P(all batteries operate correctly) = $1 - P(B_1 \text{ and } B_2 \text{ and } B_3 \text{ and } B_4) = 1 - P(B_1)P(B_2)P(B_3)P(B_4) = 1 - (.95)^4 = .1855$.

65. P(at least one event occurs) = P(A or B) = 1 - P(neither event occurs) = 1 - P(A' and B') = 1-P(A')P(B') = (1-P(A)(1-P(B)). We have used the fact that A' and B' are independent to simplify P(A' and B').

67. (a) The total area under the density curve must equal 1, so:

$$1 = \int_{-\infty}^{\infty} f(x)dx = \int_{0}^{b} \tfrac{1}{2}xdx = \tfrac{1}{2}\left[\frac{x^2}{2}\right]_0^b = \frac{b^2}{4} \text{ and, therefore, } b^2 = 4 \text{ and } b = 2.$$

(b) $$\mu = \int_{-\infty}^{\infty} xf(x)dx = \int_{0}^{b} \tfrac{1}{2}x^2dx = \tfrac{1}{2}\left[\frac{x^3}{3}\right]_0^b = \frac{b^3}{6} = \frac{2^3}{6} = \frac{4}{3}.$$

(c) $$\sigma^2 = \int_{-\infty}^{\infty} (x-\mu)^2 f(x)dx = \int_{0}^{b} (x-\tfrac{4}{3})^2 \tfrac{1}{2}xdx = \tfrac{1}{2}\int_{0}^{b} (x^2 - \tfrac{8}{3}x + \tfrac{16}{9})xdx = \tfrac{1}{2}\int_{0}^{b} (x^3 - \tfrac{8}{3}x^2 + \tfrac{16}{9}x)dx =$$

$$\tfrac{1}{2}\left[\frac{x^4}{4} - \tfrac{8}{3}\left(\frac{x^3}{3}\right) + \tfrac{16}{9}\left(\frac{x^2}{2}\right)\right]_0^b = \frac{32}{243}, \text{ so } \sigma = .36289.$$

69. Let x = the number of incorrectly filled out data fields.
x is binomial with n = 30, $\pi = 0.005$.

(a) $P(x \geq 1) = 1 - P(x=0) = 1 - (.995)^{30} = .1396$

(b) $P(x=0) = (.995)^{30} = .8604$

(c) $P(x \geq 2) = 1 - P(x \leq 1) = 1 - [P(x=0) + P(x=1)]$

$$= 1 - \left[(.995)^{30} + \binom{30}{1}(.005)^1(.995)^{29}\right]$$

$$= 1 - \left[(.995)^{30} + (30)(.005)(.995)^{29}\right] = .0099$$

71. (a)

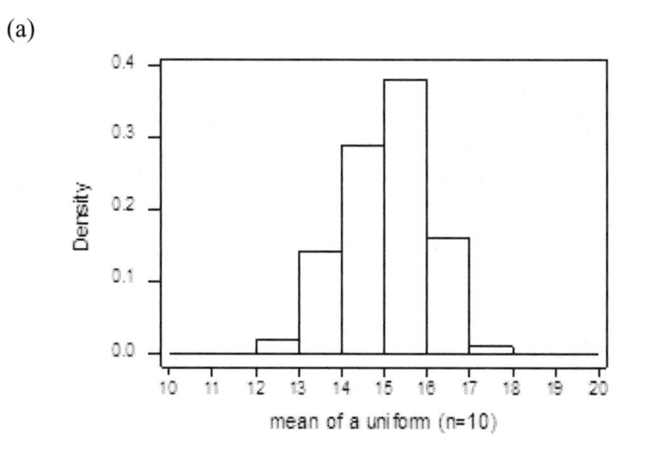

The shape of this histogram is quite symmetric and bell-shaped.

(b)

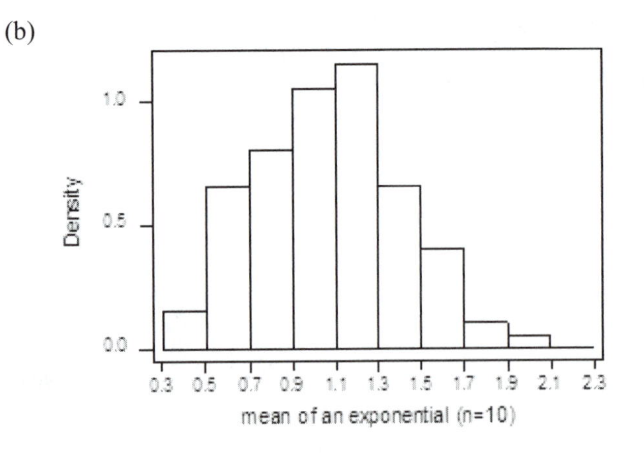

The shape of this histogram is somewhat positively skewed.

(c) As we know, averages tend to follow a normal distribution for sufficiently large n. In these cases, n = 10. When starting with a uniform distribution, a sample size of 10 is sufficiently large to produce a reasonable normal sampling distribution of $\bar{x}$. However, for the exponential distribution, a sample size of 10 is not yet sufficiently large to produce a normal sampling distribution of $\bar{x}$. The positive skewness of the exponential distribution is still evident in the sampling distribution of $\bar{x}$.

73. x = resistance n = 5 $\mu = 100$ *ohms* $\sigma = 1.7$ *ohms*

First, we must assume that the population density for x is symmetric. Then, we can use the Central Limit Theorem to proceed.

(a) Note: The question should read: "What is the probability that the *average* resistance in the circuit exceeds 105 ohms?"

$$P(\bar{x}>105)\approx P\left(\bar{x}>\frac{105-100}{1.7/\sqrt{5}}\right)=P(z>6.58)\approx 0$$

(b) $P(T>511)+P(T<489)=P\left(\bar{x}>\frac{511}{5}\right)+P\left(\bar{x}<\frac{489}{5}\right)$

$$P(\bar{x}>102.2)+P(\bar{x}<97.8)=P\left(z>\frac{102.2-100}{1.7/\sqrt{5}}\right)+P\left(z<\frac{97.8-100}{1.7/\sqrt{5}}\right)$$
$$=P(z>2.89)+P(z<-2.89)=1-P(z<2.89)+P(z<-2.89)$$
$$=1-.9981+.0019=.0038$$

(c) We know that: $P(-1.96\le z\le 1.96)=.95$

So, $\frac{\bar{x}-100}{1.7/\sqrt{n}}=1.96 \Rightarrow \bar{x}=100+1.96\left(\frac{1.7}{\sqrt{n}}\right)$

Also, $T=510=n\bar{x} \Rightarrow n=\frac{T}{\bar{x}}=\frac{510}{\bar{x}}$

Thus: $n = \dfrac{510}{100 + 1.96\left(\dfrac{1.7}{\sqrt{n}}\right)}$

Solving for n produces a value just over 5.

75. We wish to evaluate $P(1 \mid \text{late})$, $P(2 \mid \text{late})$, and $P(3 \mid \text{late})$ for both Washington, DC and Los Angeles.

First consider the flights coming into Washington, DC: Using conditional probability, we have:

$$P(1 \mid \text{late}) = \frac{P(1 \text{ and late})}{P(\text{late})} = \frac{P(\text{late and } 1)}{P(\text{late})} = \frac{P(\text{late} \mid 1)P(1)}{\sum_{i=1}^{3} P(\text{late} \mid i)P(i)}.$$

Since all posterior probabilities use the same denominator, we compute the denominator first.

$$\sum_{i=1}^{3} P(\text{late} \mid i)P(i) = P(late \mid 1)P(1) + P(late \mid 2)P(2) + P(late \mid 3)P(3)$$

$$= (.30)(.5) + (.25)(.3) + (.40)(.2) = .15 + .075 + .08 = .305.$$

So for flights coming into DC, we have:

$$P(1 \mid \text{late}) = \frac{P(\text{late} \mid 1)P(1)}{\sum_{i=1}^{3} P(\text{late} \mid i)P(i)} = \frac{.15}{.305} \approx .4918;$$

$$P(2 \mid \text{late}) = \frac{P(\text{late} \mid 2)P(2)}{\sum_{i=1}^{3} P(\text{late} \mid i)P(i)} = \frac{.075}{.305} \approx .2459;$$

$$P(3 \mid \text{late}) = \frac{P(\text{late} \mid 3)P(3)}{\sum_{i=1}^{3} P(\text{late} \mid i)P(i)} = \frac{.08}{.305} \approx .2623$$

The appropriate denominator for flights coming into Los Angeles is:

$$\sum_{i=1}^{3} P(\text{late} \mid i)P(i) = P(late \mid 1)P(1) + P(late \mid 2)P(2) + P(late \mid 3)P(3)$$

$$= (.10)(.5) + (.20)(.3) + (.25)(.2) = .05 + .06 + .05 = .16$$

So for flights coming into Los Angeles, we have:

$$P(1 \mid \text{late}) = \frac{P(\text{late} \mid 1)P(1)}{\sum_{i=1}^{3} P(\text{late} \mid i)P(i)} = \frac{.05}{.16} = .3125;$$

$$P(2 \mid \text{late}) = \frac{P(\text{late} \mid 2)P(2)}{\sum_{i=1}^{3} P(\text{late} \mid i)P(i)} = \frac{.06}{.16} = .375$$

$$P(3 \mid \text{late}) = \frac{P(\text{late} \mid 3)P(3)}{\sum_{i=1}^{3} P(\text{late} \mid i)P(i)} = \frac{.05}{.16} = .3125$$

77. By the addition rule, we have .626 = P(A or B) = P(A) + P(B) – P(A and B).
So .626 = P(A) + P(B) - .144, from which it follows that .770 = P(A) + P(B). (*)

Now, since A and B are independent, it follows that .144 = P(A and B) = P(A)P(B).
So P(A) = .144/P(B). Substitute this into the (*) above to obtain:

$$.770 = \frac{.144}{P(B)} + P(B) \Rightarrow .770P(B) = .144 + [P(B)]^2.$$

Solving the quadratic equation $[P(B)]^2 - .770P(B) + .144 = 0$ yields the solution .45 and .32.
If P(B) = .45, then P(A) = .144/.45 = .32, which contradicts the fact that P(A) > P(B).

So it must be the case that P(B) = .32. Then P(A) = .144/.32 = .45.
So the solutions are P(A) = .45 and P(B) = .32.

Chapter 6

Quality and Reliability

Section 6.1

1. The specification limits are 5% above and below the nominal value of 560. This tolerance is $\pm(.05)(560) = \pm 28$ Ohms, so LSL = 560-28 = 532Ω and USL = 560 + 28 = 588Ω.

3. (a) The envelope puts an upper specification limit of 4.00 inches on the width of a folded letter.

   (b) Possible penalties: you may have to refold letter (rework), or bend letter to fit envelope (lower quality), or reprint and fold new letter (scrap and rework).

5. (a) attributes data
   (b) variables data
   (c) variables data
   (d) attributes data
   (e) attributes data
   (f) variables data
   (g) attributes data
   (h) variables data
   (i) variables data

   One often-used rule for deciding whether a variable is continuous is the following: *if it is possible to obtain more and more precise measured values by using better and better measuring instruments, then the characteristic/variable is continuous (i.e, variables data).* Note that you don't have to actually obtain and use better instruments, you just have to perform the thought experiment of asking if instruments exist that will conceivably give more and more precise measurements.

7. Some unacceptable parts whose *true* lengths are .02 inches or less below the LSL will give measured lengths above the LSL (and will then be incorrectly classified as acceptable). Conversely, some acceptable parts whose true lengths are less than .02 inches below the USL will have measured lengths above the USL (which incorrectly classifies them as unacceptable).

Section 6.2

9. Method 2 would be a better rational subgrouping scheme. If it is believed that the machine used to produce the part may be a source of special cause variation, subgroups should be created which are machine specific. In this way each machine's output can be assessed. Method 1 does not allow for the identification of the machine, nor does it create subgroups that are machine specific.

11. (a) $P(z > 3) = .0013$
   $\Rightarrow$ The probability that a single control chart point falls above the UCL in a 3-sigma control chart is .0013.

   (b) $P(z > 3.09) = .001$
   $\Rightarrow$ The probability that a single control chart point falls above the UCL in a 3.09-sigma control chart is .001.

13. Chart #1: This process is out of statistical control. Test #3 is found [Six points in a row are steadily increasing, starting with point #3.]

Chart #2: Even though there are no Tests found, Test #7 (which requires that 15 points in a row be in zone C) seems likely to occur. Chart #2 only has 13 points, but all of them appear to be in zone C. So, while the process is currently in statistical control, with more process data it seems likely to go out of statistical control.

Chart #3: This process is out of statistical control. Test #2 is found [Nine points in a row on one side of the centerline, starting with point #2.]

Chart #4: This process is out of statistical control. Both Tests #5 and 6 are found starting with point #1. The process values begin beyond zone A and remain there for 5 consecutive points.

Section 6.3

15. Given: $\sum R = 85.2 \quad k = 30 \quad n = 4$

Centerline $\overline{R} = \left(\dfrac{85.2}{30}\right) = 2.84$

$UCL_R = D_4\overline{R} = (2.282)(2.84) = 6.48$

$LCL_R = D_3\overline{R} = (0)(2.84) = 0$

(Note: D_4 and D_3 were obtained from Table XI; Control Chart Constants)

17. (a)

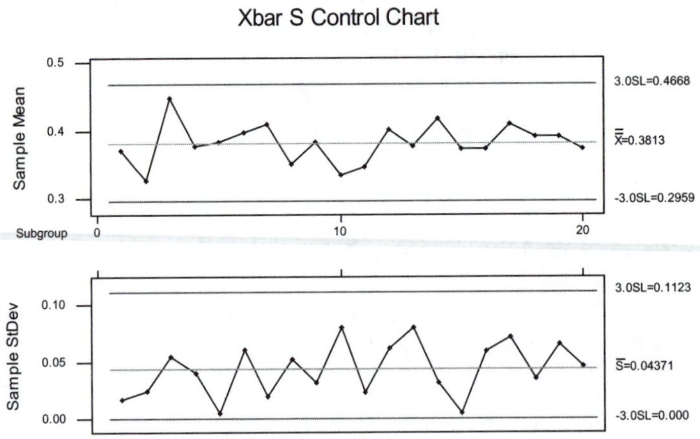

The above $\overline{x}$, s chart was constructed in Minitab.

On the s chart no rules for statistical control are broken. So, we would conclude that the process variation is in statistical control.

(b) On the $\overline{x}$ chart there are no 'out of control' conditions. So, the process is in statistical control.

Notice that the control limits on the $\overline{x}$ chart are the same as those on the $\overline{x}$ chart in Exercise 16(b). Consider the methods used to obtain the control limits in both exercises:

Exercise #16(b) $\overline{\overline{x}} \pm A_2\overline{R} \Rightarrow .3813 \pm (1.023)(.0835)$

So, $UCL_{\overline{x}} = .4667,\ LCL_{\overline{x}} = .2959$

Exercise #17(b) $\overline{\overline{x}} \pm A_3\overline{s} \Rightarrow .3813 \pm (1.954)(.04371)$

So, $UCL_{\overline{x}} = .4667,\ LCL_{\overline{x}} = .2959$

Exercises #16 and 17 illustrate that regardless of the method used to estimate the process variation (R or s) the control limits for the $\overline{x}$ chart, which are set at $\pm 3\frac{\sigma}{\sqrt{n}}$, will be essentially the same.

One should be aware, however, that since two different methods of estimating the variation may be used, the $\overline{x}$ control limits may vary a little. We just did not see much of a difference in this exercise.

19. (a) A Minitab *s* chart of this data appears below. The centerline is 1.2642, UCL = $B_4\overline{s}$ = 1.970(1.2642) = 2.4905, and LCL = $B_3\overline{s}$ = (0.030)(1.2642) = 0.0379. *Note: to use Minitab to create this chart, you must first create 24 subgroups of size 6 whose subgroup means and standard deviations match those given in the exercise. An easy way to do this is to take any 6 numbers, store them in a Minitab column, say C50, and then type the Minitab expression: MTB> let c1 =c51(1)+c52(1)*((c50-mean(c50))/stdev(c50)). Here, c51and c52 are columns containing the means and standard deviations given in the exercise. The data in column c1 will then have a mean and standard deviation that exactly matches that of the first subgroup. Repeat this procedure (use Minitab's Command Editor) for creating the remaining subgroups and store them in columns c2 through c24.*

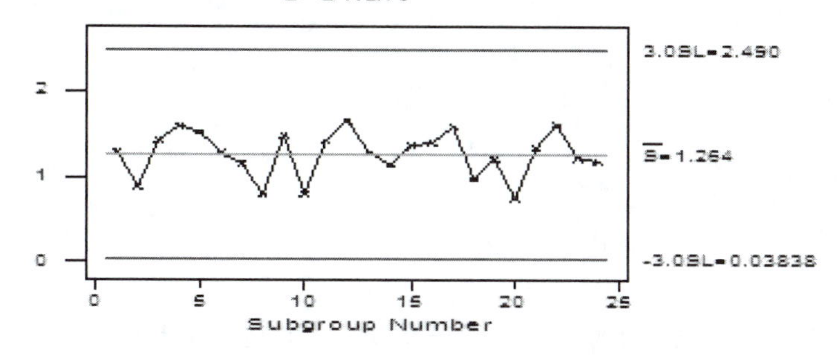

(b) A Minitab $\overline{x}$ chart of this data appears below. The centerline is 96.503. The control limits, based on the s chart centerline, are UCL = $\overline{\overline{x}} + A_3\overline{s}$ = 96.503+1.287(1.2642) = 98.1300 and LCL = $\overline{\overline{x}} - A_3\overline{s}$ = 96.503-1.287(1.2642) = 94.8760.

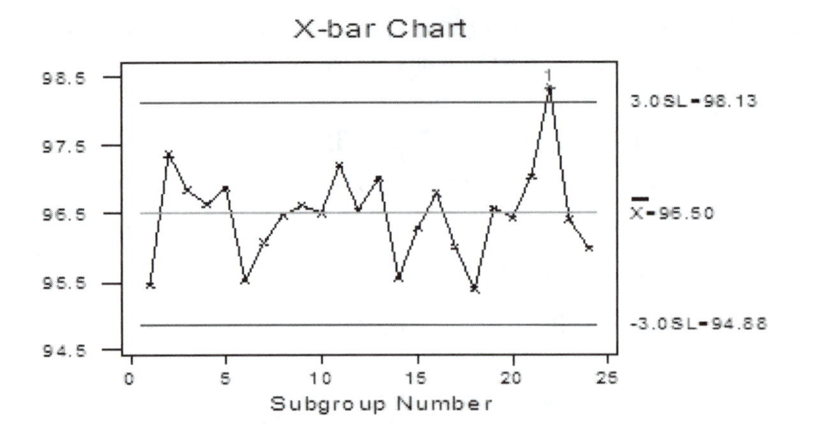

21. (a) If each x_i value is transformed into $y_i = b(x_i - a)$, where a and b are constants and b > 0, then for any set of n values, $\overline{y} = b(\overline{x} - a)$ and $\overline{R_y} = b\overline{R_x}$. From these two relationships, simple algebra will show that, for example, x_i > UCL (of the x-data) if and only if y_i > UCL (of the y-data):

$x_i > UCL_x = \overline{\overline{x}} + A_2\overline{R_x} \Rightarrow y_i = b(x_i - a) > b(\overline{\overline{x}} - a + A_2\overline{R_x}) = b(\overline{\overline{x}} - a) + A_2(b\overline{R_x}) = \overline{\overline{y}} + A_2\overline{R_y} = UCL_y$ That is, the x-bar charts based on un-transformed data and transformed data will give the same signals. In the same manner, it can be shown that the R charts for both transformed and un-transformed data give the same signals.

(b) The transformed data is shown below:

i	y_1	y_2	y_3	i	y_1	y_2	y_3
1	4	0	2	11	-1	3	0
2	-1	-3	-1	12	-2	-1	4
3	-2	4	2	13	4	-1	3
4	-2	-2	1	14	-3	3	2
5	0	-2	2	15	2	0	3
6	-1	0	2	16	-3	1	-1
7	-3	3	3	17	-2	2	1
8	-2	-3	1	18	-3	2	-1
9	-3	1	3	19	-1	-2	0
10	3	1	1	20	1	-2	-1

The centerline of the R chart is 4.100. The UCL = $D_4\overline{R}$ = (2.574)(4.10) = 10.5534. The LCL = $D_3\overline{R}$ = (0)(4.10) = 0.0000. The centerline of the x-bar chart is 0.2500. The UCL = $\overline{\overline{x}} + A_2\overline{R}$ = .25000+(1.023)(4.10) = 4.444. And, the LCL = $\overline{\overline{x}} - A_2\overline{R}$ = .25000-(1.023)(4.10) = -3.944. There are no 'out of control' conditions in either chart.

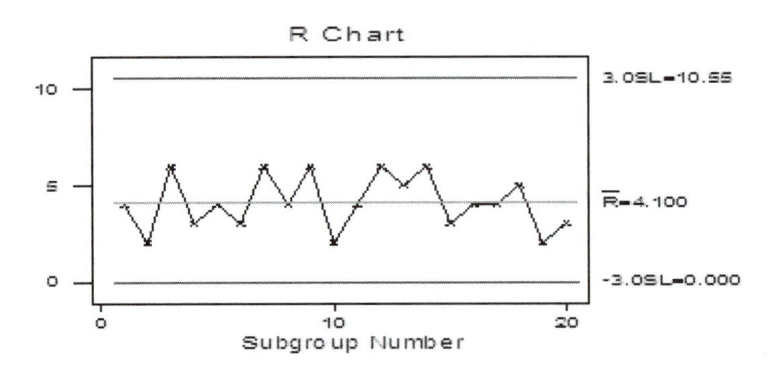

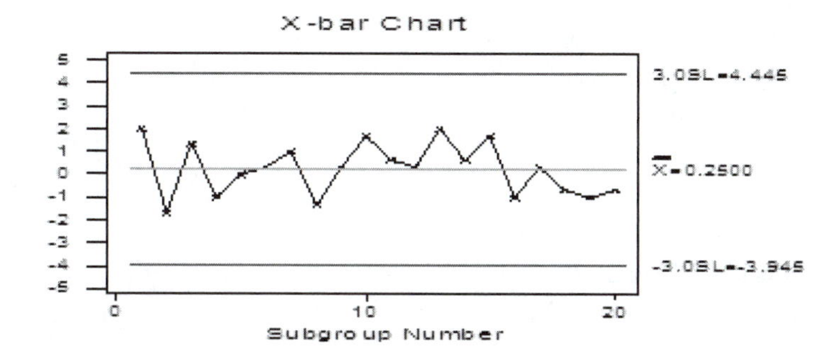

(c) The centerline of the R chart is 0.0041, UCL = $D_4 \overline{R}$ = (2.574)(.0041) = .010553, LCL = $D_3 \overline{R}$ = (0)(.0041) = 0.0000. The centerline of the x-bar chart is 0.2542 , UCL = $\overline{\overline{x}} + A_2 \overline{R}$ = .2542+(1.023)(.0041) =.2584, and LCL = $\overline{\overline{x}} - A_2 \overline{R}$ = .2542-(1.023)(.0041) = 0.2501. There are no 'out of control' conditions in either chart.

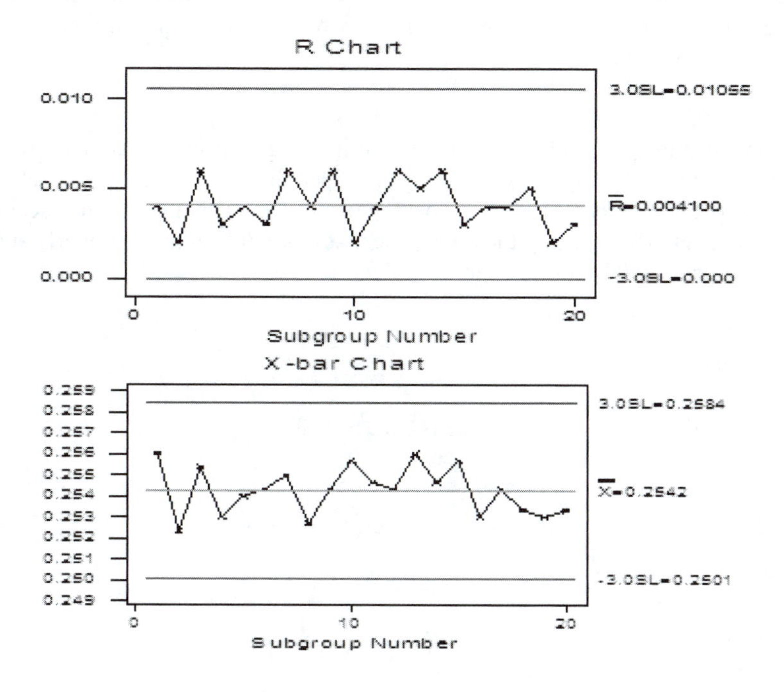

23. (a) An estimate of the process standard is $\hat{\sigma} = \left(\frac{\overline{s}}{c_4}\right)$

$$\Rightarrow \hat{\sigma} = \left(\frac{.04371}{.8862}\right) = .04932$$

[Alternatively, $\hat{\sigma} = \frac{\overline{R}}{d_2} = \left(\frac{.0835}{1.693}\right) = .04932$]

(b) P(x > USL) = P(x > .48)

$$=P\left(z>\frac{.48-.3813}{.04932}\right)=P(z>2.00)=.0228$$

P(x < LSL) = P(x < .32)

$$=P\left(z<\frac{.32-.3813}{.04932}\right)=P(z<-1.24)=.1075$$

Section 6.4

25. If a process is in a state of statistical control it means that it is stable and predictable. In such a case it makes sense to compare the process output to the process specifications. However, if a process is not in statistical control, then we do not have stable output and so there is no use in comparing this output to the specifications. After all, any comparisons done in this case may be unreliable since our process output may be changing in an unpredictable manner.

27. Since the $C_p > 1$ we know that the process has the *potential* of meeting both specifications. However, since the $C_{pk} < 1$, the process is not actually capable of meeting both specifications. With a $C_{pk} = .9$ and a $C_p = 1.6$ we can determine that one of the specifications is met but one of them is not met. The specification that is not met is just barely not met. By adjusting the process so that it is centered more properly at the center of the specifications, the process will become capable.

29. *Proportion out of specification*

$$= P(z\geq(3)(.9))+P(z\geq[(6)(1.6)-(3)(.9)])=P(z\geq2.7)+P(z\geq6.9)=.0035$$

31. $\bar{x}=2.5033 \quad USL=2.55$

$s=.021227 \quad LSL=2.45$

$$C_p=\left(\frac{2.55-2.45}{(6)(.021227)}\right)=.785$$

$$C_{pu}=\left(\frac{2.55-2.5033}{(3)(.021227)}\right)=.733$$

$$C_{p\ell}=\left(\frac{2.5033-2.45}{(3)(.021227)}\right)=.837$$

$$C_{pk}=\min(.733,.837)=.733$$

33. (a) To compute capability indexes on the transformed process data, the control chart statistics from the transformed data should be used.

So, $\bar{\bar{x}}=.1833$, $\bar{R}=4.200$. Also, the process specifications need to be transformed. So, USL = 10 and LSL = -10. Using these values the C_p indexes can be computed for the transformed data.

(b) $\hat{\sigma}=\left(\frac{\overline{R}}{d_2}\right)=\left(\frac{4.200}{1.693}\right)=2.48$

$C_p=\left(\frac{10-(-10)}{(6)(2.48)}\right)=1.34$

$C_{pu}=\left(\frac{10-(.1833)}{(3)(2.48)}\right)=1.32$

$C_{p\ell}=\left(\frac{.1833-(-10)}{(3)(2.48)}\right)=1.37$

$C_{pk}=\min(1.32,1.37)=1.32$

This process is capable.

35. $\hat{\sigma}=\overline{R}/d_2 = .01531/1.128 = 0.013573$

C_p = (USL-LSL)/ 6$\hat{\sigma}$ = 0.8/(6*.013573) = 9.824

$\hat{\mu}$ = 59.66828

k = (60 – 59.66828)/(.4) = 0.8293

C_{pk} = (1-.8293)(9.824) = 1.677

Because both C_p and C_{pk} are greater than 1, this implies that the process is capable of staying within the specification limits.

Section 6.5

37. $LCL=\left(\overline{p}-3\sqrt{\frac{\overline{p}(1-\overline{p})}{n}}\right)=0$

$\Rightarrow \overline{p}=3\sqrt{\frac{\overline{p}(1-\overline{p})}{n}}$

$\Rightarrow \overline{p}^2=\frac{9\overline{p}(1-\overline{p})}{n}$

$\Rightarrow n\overline{p}^2=9\overline{p}-9\overline{p}^2$

$\Rightarrow (n+9)\overline{p}^2=9\overline{p}$

$\Rightarrow \overline{p}=\left[\frac{9}{n+9}\right]$

39. (a)

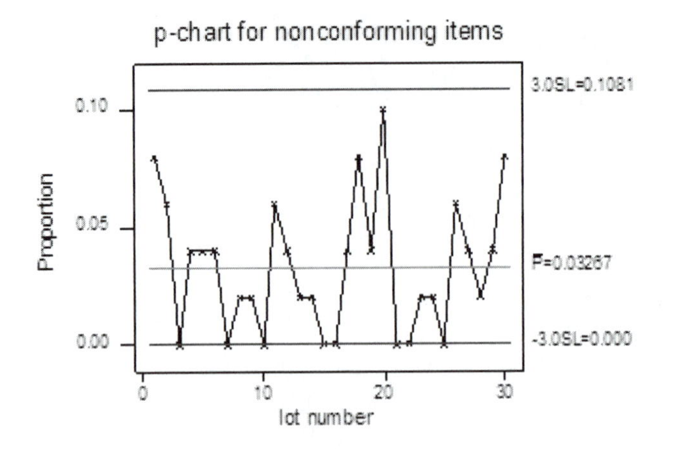

Computations: $\bar{p} = \left(\frac{49}{1500}\right) = .032667$

$$UCL = .032667 + 3\sqrt{\frac{(.032667)(.96733)}{50}} = .1081$$

$$LCL = .032667 - 3\sqrt{\frac{(.032667)(.96733)}{50}} = -.042752$$

However, because $\bar{p}$ is so small (which produces a negative lower control limit), we will replace the lower control limit by 0, since it is impossible to have negative nonconformance rates.

(Note: Many practitioners require that n and $\bar{p}$ be sufficiently large to produce a nonnegative LCL.)

(b) There are 8 lots which contain no nonconforming items. These values fall on the LCL, however, in this case we should probably not be too surprised by this. Since there are no signs of any 'out of control' conditions, we conclude that the process is in statistical control and currently operating at about a 3.3% nonconforming rate.

41. (a)

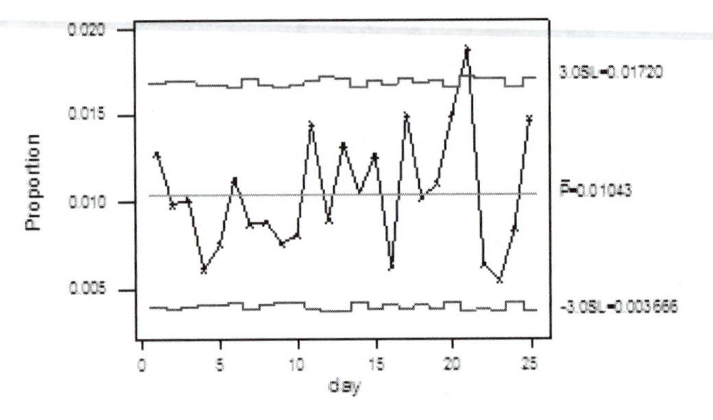

$$\overline{p}=\left(\frac{571}{54,726}\right)=0.01043$$

(b) As shown on the control chart, the control limits vary with each day since the number of keyboards tested each day varies.

The general formulas are:

$$UCL=.01043+3\sqrt{\frac{(.01043)(.98957)}{n_i}}$$

$$LCL=.01043-3\sqrt{\frac{(.01043)(.98957)}{n_i}}$$

The control limits listed on the control chart correspond to the smallest number of keyboards per day (n = 2,011).

(c) On day 21, the proportion of keyboards failing inspection is .0189. This value is above the upper control limit. The production process on this day should be studied to see if an assignable cause can be found. Appropriate actions should be taken to prevent a reoccurrence of the problem.

No other 'out of control' conditions are present.

43.

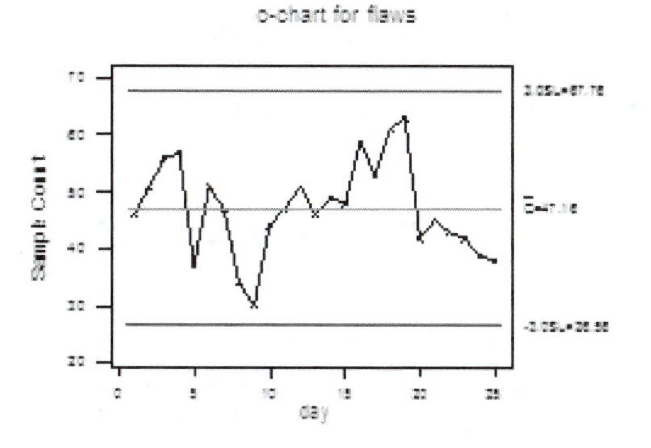

A c-chart is required in this case:

Computations are: $\overline{c}=\left(\frac{1,179}{25}\right)=47.16$

$$UCL=47.16+3\sqrt{47.16}=67.76$$

$$LCL=47.16-3\sqrt{47.16}=26.56$$

When analyzing the control chart we do not see any 'out of control' conditions. However, the last 6 days of production have produced below average numbers of flaws. The engineer should keep her eye on the process so that if the process continues in this manner, she can investigate what assignable cause may have improved the process. Hopefully, the process will continue to function in this improved manner.

45. (a)

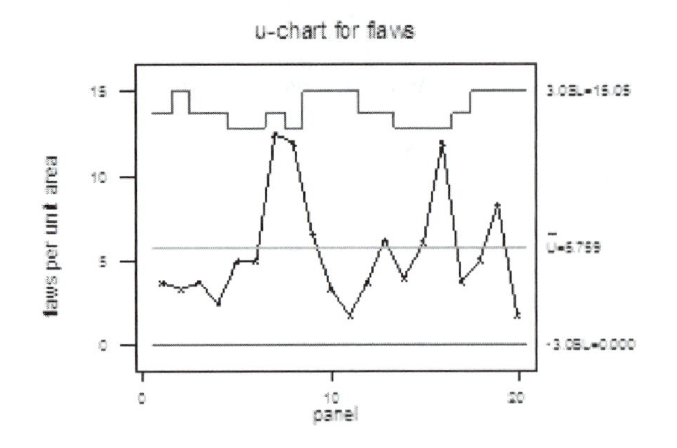

Computations: $\bar{u} = \left(\frac{91}{15.8}\right) = 5.759$

As shown on the control chart, the control limits vary with each panel since the panel areas vary. The general formulas are:

$$UCL = 5.759 + 3\sqrt{\frac{5.759}{n_i}}$$

$$LCL = 5.759 - 3\sqrt{\frac{5.759}{n_i}}$$

The control limits listed on the control chart correspond to the smallest panel area (n = .6).

Notice that the lower control limit is negative but that it has been replaced by 0, since it is impossible to have negative flaw rates.

(Note: As stated earlier, many practitioners require that n and $\bar{u}$ be sufficiently large to produce a nonnegative LCL.)

(b) Panels #7 and 8 seem to have significantly larger flaw rates than the process average. Test #5 (two out of three points in a row outside of 2 standard deviations) is observed. Also, panel #16 has the same flaw rate as panel #8, though no rules of statistical control are broken. Overall, it would be prudent for the engineer to investigate the process that was operating when these panels were produced in order to detect an assignable cause and, hopefully, prevent it from reoccurring.

Section 6.6

47. (a) Let x denote the number of cycles the spring can be compressed and released. Because x follows a Weibull distribution, the reliability is given by $R(t) = \int_t^{\infty} \frac{\alpha}{\beta^\alpha} x^{\alpha-1} e^{-(x/\beta)^\alpha} dx = e^{-(t/\beta)^\alpha}$. Substituting for $t = 400{,}000$ and the parameters $\alpha = 4$ and $\beta = 600{,}000$ gives $R(400{,}000) = e^{-(400{,}000/600{,}000)^4} = .82075$

(b) $R(800{,}000) = e^{-(800{,}000/600{,}000)^4} = .0424$

(c) $R(600{,}000) = e^{-(600{,}000/600{,}000)^4} = 1/e = .3679$

(d) The hazard function (or failure rate) is given by $Z(t) = f(t)/R(t)$. For the Weibull distribution, the hazard function is

$$Z(t) = \frac{(\alpha/\beta^{\alpha})t^{\alpha-1}\exp\{-(t/\beta)^{\alpha}\}}{\exp\{-(t/\beta)^{\alpha}\}} = \frac{\alpha}{\beta^{\alpha}}t^{\alpha-1}.$$

For this particular mechanical spring, the hazard function is $(4/600{,}000^4)t^3$. To determine whether $Z(t)$ is increasing, decreasing, or flat, take the first derivative of $Z(t)$ with respect to *t*. Doing so yields $dZ/dt = \frac{\alpha}{\beta^{\alpha}}(\alpha-1)t^{\alpha-2}$. Substituting the appropriate parameter values for this problem gives $dZ/dt = \frac{(4)(3)}{600{,}000^4}t^2$, which is positive for all $t > 0$. Hence, this failure function is **increasing.**

49. (a) The following graph depicts the failure rate for a product that is normally distributed with $\mu = 10$ and $\sigma = 2$.

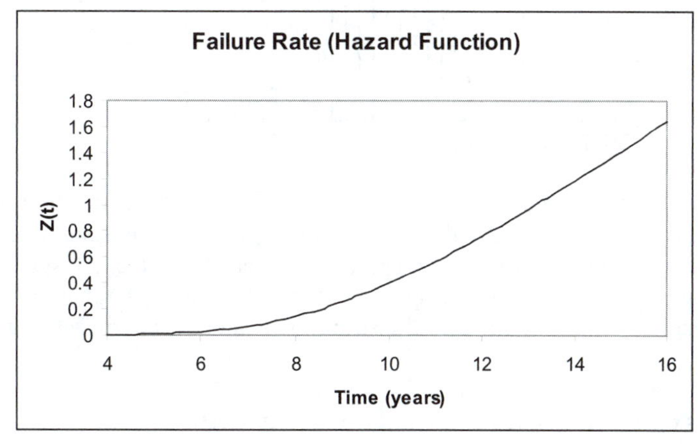

(b) It can be seen from the graph in part (a) that normal failure laws have an **increasing** rate.

51. (a) The following diagram depicts this particular RAID system. Here, the subscript "m" refers to the irror disk.

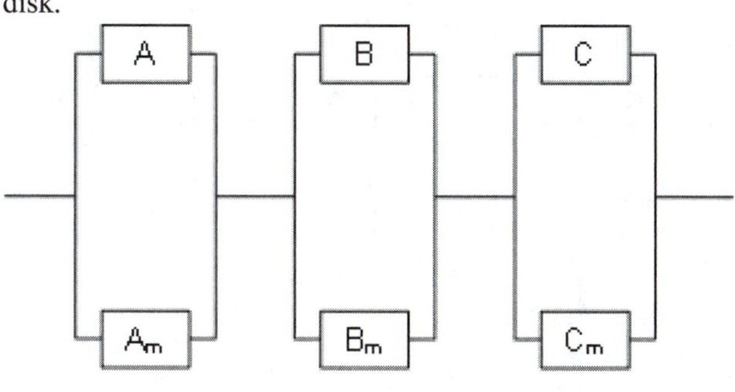

(b) Let R_1, R_2, and R_3 refer to the reliabilities of pairs A, B, and C, respectively. Since the overall system is comprised of the pairs connected in series, the overall reliability is given by

$R(t) = R_1(t) \cdot R_2(t) \cdot R_3(t)$. Clearly, $R_1(t) = R_2(t) = R_3(t)$. Because pair A is connected in parallel, we have $R_1(t) = 1 - [1 - R_A(t)] \cdot [1 - R_{A_m}(t)]$. Using the fact that any disk has an exponential lifetime with parameter $\lambda = .025$, we have $R_A(t) = R_{A_m}(t) = e^{-\lambda t} = e^{-.025t}$. Therefore, $R_1(t) = (1 - [1 - e^{-.025t}] \cdot [1 - e^{-.025t}])$, and so $R(t) = [R_1(t)]^3 = (1 - [1 - e^{-.025t}] \cdot [1 - e^{-.025t}])^3$.

Supplementary Exercises

53.

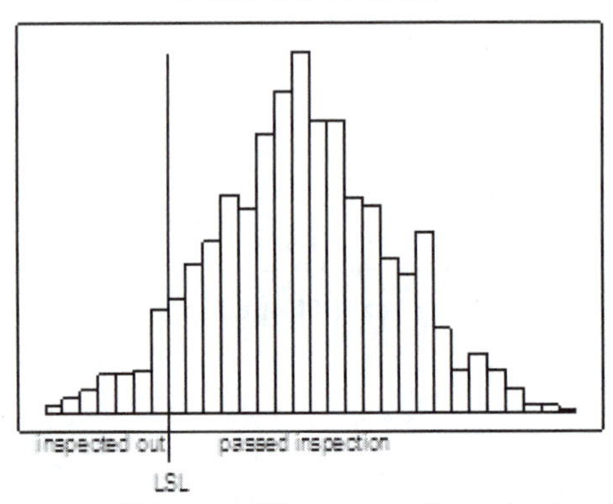

Bottles that pass inspection will have a fill volume of at least the fill specification (LSL). The maximum fill volume is determined by the size of the bottle. (Although this value is not likely reached.) The shape of the distribution of fill volumes that pass inspection is truncated on the left, since bottles with fill volumes below the lower specification have been inspected out, resulting in the left part of the distribution being "cut-off". The histogram illustrates a normal distribution that has been truncated at the LSL.

55. As the drill wears out it may not be able to drill the hole diameters properly. On a control chart this problem will likely manifest itself as a slow trend down in the hole diameters that are being drilled. That is, the hole diameters may get smaller and smaller. Test #3 on the conditions for an 'out of control' process may occur.

57. (a) The transformed data was achieved as in Exercise 21.

Fore example, the first observation in subgroup 1, $x_1 = .251$ was transformed as follows: [(.251 - .250)(1000)] = 1, where .250 was the nominal value for a bar of type P2.

Subgroup	Nominal	y1	y2	y3	y4
1	0.250	1	2	0	-1
2	0.375	-3	3	4	0
3	0.250	-3	-1	4	1
4	0.250	-2	-3	0	2
5	0.250	-1	-1	0	-1
6	0.125	0	2	0	1
7	0.375	-3	-1	0	1
8	0.500	-1	2	-5	3
9	0.125	-1	-4	-2	1
10	0.125	1	1	5	-3
11	0.375	0	-1	3	4
12	0.250	-1	-1	0	-3
13	0.250	0	3	1	-2
14	0.250	-1	0	-1	-1
15	0.250	2	0	1	-3

16	0.250	1	-1	0	0
17	0.125	1	2	-3	0
18	0.125	-2	-2	-2	3
19	0.250	2	0	-3	-2
20	0.500	2	-4	2	2

(b)

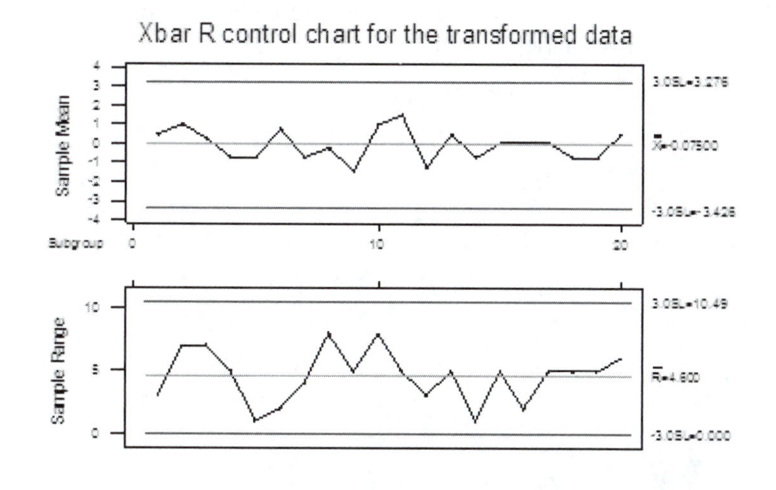

When analyzing the control chart we do not see any 'out of control' conditions. We conclude that the milling process is in statistical control. The only issue an engineer might address is that the process mean is –.07500. This means that, on average, the bars are being milled a bit under the nominal values.

59. $\hat{\sigma} = \left(\frac{\overline{R}}{d_2}\right) = \left(\frac{1.163}{2.326}\right) = .5$

where d_2 for a subgroup size of 5 is found on Table XI: Control Chart Constants.

$$C_p = \left(\frac{16-12}{(6)(.5)}\right) = 1.33$$

$$C_{pu} = \left(\frac{16-14.5}{(3)(.5)}\right) = 1$$

$$C_{pl} = \left(\frac{14.5-12}{(3)(.5)}\right) = 1.67$$

$$C_{pk} = \min(1, 1.67) = 1$$

61. Let x have an exponential distribution with parameter λ. So:

$$P(x > t_1 + t_2 \mid x > t_1) = \frac{P(x > t_1 + t_2 \ \ and \ \ x > t_1)}{P(x > t_1)} = \frac{P(x > t_1 + t_2)}{P(x > t_1)}$$

$$= \frac{\int_{t_1+t_2}^{\infty} \lambda e^{-\lambda x} dx}{\int_{t_1}^{\infty} \lambda e^{-\lambda x} dx} \text{ (or } R(t_1+t_2)/R(t_1)\text{)} = e^{-(t_1+t_2)\lambda} / e^{-t_1\lambda} = e^{-t_2\lambda} = P(x > t_2)$$

63. (a) Since this system is connected in series, the overall reliability is $R(t) = R_1(t) \cdot R_2(t)$. Note that $R_1(t) \le 1$ and $R_2(t) \le 1$, and so $R(t) \le R_1(t)$ and $R(t) \le R_2(t)$. Thus, the overall reliability never exceeds the reliability of any of its individual components. That is, $R(t) \le \min\{R_1(t), R_2(t)\}$.

(b) In the case where the components are not necessarily independent, then $R(t) = P(T > t)$ = P(both components last longer than t) = $P(T_1 > t \text{ and } T_2 > t)$. Since $\{T_1 > t \text{ and } T_2 > t\}$ is the intersection of the two events $\{T_1 > t\}$ and $\{T_2 > t\}$, it's probability cannot exceed $P(T_1 > t)$ or $P(T_2 > t)$. That is, $R(t) \le \min[P(T_1 > t), P(T_2 > t)] = \min[R_1(t), R_2(t)]$.

Chapter 7

Estimation and Statistical Intervals

Section 7.1

1. A single randomly selected item forms a sample of size n = 1. If x denotes the value of this item, then the sample average is also equal to x. Since the sample average, based on any sample size, is an unbiased estimator or the population mean, then the length x is indeed an unbiased estimator of the population mean. Of course, this estimator is not as precise as the sample average based on larger sample sizes (i.e., it has a larger standard error), but it is still unbiased.

3. (a) Because the population is known to be normal and its standard deviation is known to be σ = 5, the sampling distribution of $\bar{x}$ is also normal. Standardizing, $P(\mu-1 \le \bar{x} \le \mu+1) = P(\frac{(\mu-1)-\mu}{\sigma/\sqrt{n}} \le z \le \frac{(\mu+1)-\mu}{\sigma/\sqrt{n}}) = P(-\frac{\sqrt{n}}{\sigma} \le z \le \frac{\sqrt{n}}{\sigma}) = P(-.63 \le z \le .63) = .7357 - .2643 = .4714$.

(b) In each case, as in part (a), the desired probability is $P(-\frac{\sqrt{n}}{\sigma} \le z \le \frac{\sqrt{n}}{\sigma})$:

For n = 50,

$P(-\frac{\sqrt{n}}{\sigma} \le z \le \frac{\sqrt{n}}{\sigma}) = P(-1.41 \le z \le 1.41) = .9207 - .0793 = .8414$.

For n = 100,

$P(-\frac{\sqrt{n}}{\sigma} \le z \le \frac{\sqrt{n}}{\sigma}) = P(-2.00 \le z \le 2.00) = .9772 - .0228 = .9544$.

For n = 1000,

$P(-\frac{\sqrt{n}}{\sigma} \le z \le \frac{\sqrt{n}}{\sigma}) = P(-6.32 \le z \le 6.32) \approx 1.0000 - .0000 = 1.0000$.

In other words, as n increases, it becomes more and more likely that the sample mean will fall within ±1 standard deviation from the population mean.

(c) As the following graph shows, the probability that the sample mean lies within ±1 unit of μ increases as the sample size n increases:

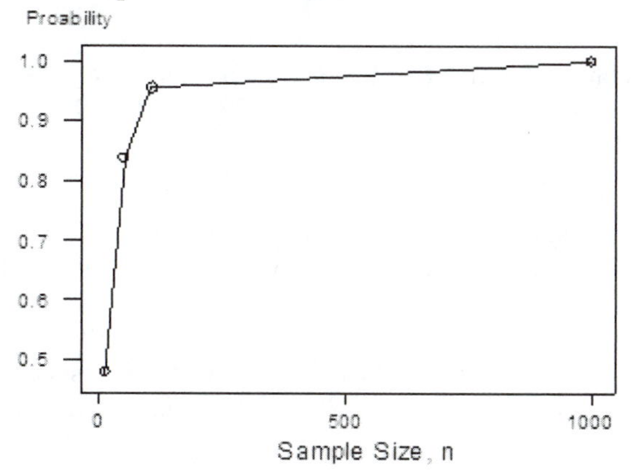

5. (a) As in exercise 3, we have P(μ-1 ≤ $\bar{x}$ ≤ μ + 1) = P($\frac{(\mu-1)-\mu}{\sigma/\sqrt{n}} \le z \le \frac{(\mu+1)-\mu}{\sigma/\sqrt{n}}$) = P(-$\frac{\sqrt{n}}{\sigma}$ ≤ z ≤ $\frac{\sqrt{n}}{\sigma}$). In this exercise, σ = 2. To make this probability equal to 90%, first use Table I (pages 534-535) to find the (approximate) value z = 1.645 for which P(-1.645 ≤ z ≤ 1.645) ≈ .90. Then equate 1.645 and $\frac{\sqrt{n}}{\sigma}$ and solve for n: $\frac{\sqrt{n}}{\sigma}$ = 1.645, so $\sqrt{n}$ =1.645(σ) = 1.645(2) and, squaring, n = (4)(1.645)^2 = 10.82. Rounding off to the nearest integer value, n ≥ 11 will guarantee that there is at least a 90% probability that the sample mean (based on a sample of size 11) will fall within ±1 standard deviation of the population mean.

(b) Using Table I : P(-1.28 ≤ z ≤ 1.28) ≈ 80%. Therefore, $\frac{\sqrt{n}}{\sigma}$ = 1.28, so $\sqrt{n}$ =1.28(σ) = 1.28(2) and n = 4(1.28)^2 = 6.55, so n ≥ 7. Similarly, P(-1.96 ≤ z ≤ 1.26) ≈ 95%. Therefore, $\frac{\sqrt{n}}{\sigma}$ = 1.96, so $\sqrt{n}$ =1.96(σ) = 1.96(2) and n = 4(1.96)^2 = 15.4, so n ≥ 16. *Note: in every case we round up to the next higher integer, because we want to be sure we have exceeded the specified probability requirement.* Finally, P(-2.575 ≤ z ≤ 2.575) ≈ 99%. Therefore, $\frac{\sqrt{n}}{\sigma}$ = 2.575, so $\sqrt{n}$ =2.575(σ) = 2.575(2) and n = 4(2.575)^2 = 26.52, so n ≥ 27.

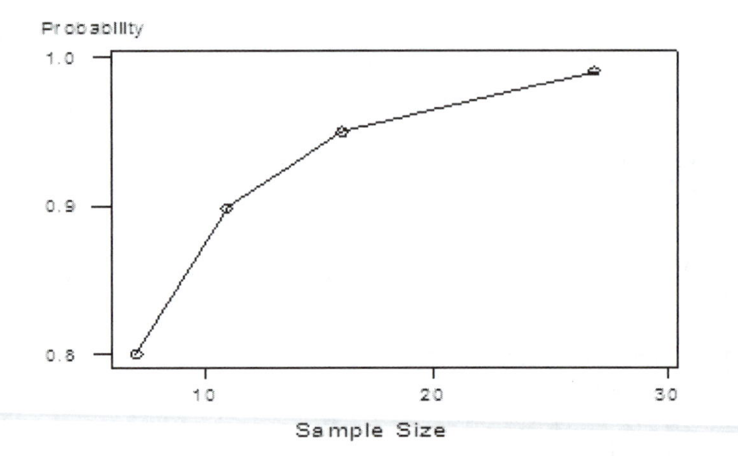

Section 7.2

7. (a) The entry in the 3.0 row and .09 column of the z table is .9990. Similarly, the entry for -3.09 is .0010. Therefore, the area under the z curve between -3.09 and +3.09 is .9990 - .0010 = .9980. The confidence level is then 99.8%.

(b) Following the example in part (a), the z-table entries corresponding to z = -2.81 and z = +2.81 are .9975 and .0025, respectively. Therefore the area between these two z values is .9975 - .0025 = .9950. The confidence level is then 99.5%.

(c) The z-table entries corresponding to z = -1.4 and z = +1.44 are .9251 and .0749, respectively. Therefore, the area between these two z values is .9251 - .0749 = .8502. The confidence level is then 85.02%.

(d) The coefficient of $s/\sqrt{n}$ is not written, but is understood to be 1.00. The z-table entries corresponding to z = -1.001 and z = +1.00are .8413 and .1587, respectively. Therefore, the area between these two z values is .8413 - .1587 = .6826. The confidence level is then 68.26%.

9. (a) The larger the confidence level, the larger the z value. Since the plus-or-minus width of the confidence interval equals $z\frac{s}{\sqrt{n}}$, the larger z values will result in wider confidence intervals. (Conversely, smaller confidence levels will result in narrower confidence intervals.)

(b) Because n appears in the denominator of the expression $z\frac{s}{\sqrt{n}}$, larger values of n will result in narrower confidence intervals. (Conversely, smaller sample sizes will result in wider confidence intervals.)

(c) Because s appears in the numerator of the expression $z\frac{s}{\sqrt{n}}$, larger values of s will result in wider confidence intervals. (Conversely, smaller values of s will result in narrower confidence intervals.)

11. (a) Decreasing the confidence level from 95% to 90% will decrease the associated z value and therefore make the 90% interval narrower than the 95% interval. *(Note: see the answer to Exercise 9 above)*

(b) The statement is not correct. Once a particular confidence interval has been created/calculated, then the true mean is either in the interval or not. The 95% refers to the process of creating confidence intervals; i.e., it means that 95% of all the possible confidence intervals you could create (each based on a new random sample of size n) will contain the population mean (and 5% will not).

(c) The statement is not correct. A confidence interval states where plausible values of the population mean are, not where the individual data values lie. In statistical inference, there are three types of intervals: **confidence intervals** (which estimate where a population mean is), **prediction intervals** (which estimate where a single value in a population is likely to be), and **tolerance intervals** (which estimate the likely range of values of the items in a population. The statement in this exercise refers to the likely range of all the values in the population, so it is referring to a tolerance interval, not a confidence interval.

(d) No, the statement is not exactly correct, but it is close. We *expect* 95% of the intervals we construct to contain μ, but we also expect a little variation. That is, in any group of 100 samples, it is possible to find only, say, 92 that contain μ. In another group of 100 samples, we might find 97 that contain μ, and so forth. So, the 95% refers to the *long run* percentage of intervals that will contain the mean. 100 samples/intervals is not the long run.

13. (a) $\bar{x} \pm 2.576\frac{s}{\sqrt{n}} = 13.83 \pm 2.576\frac{5.05}{\sqrt{131}} = (12.68, 14.98)$. We are 99% confident the average backpack weight of all 6th graders is between 12.68lbs and 14.98lbs.

(b) $\bar{x}$ for sample mean weight as a percentage of body weight is the midpoint of the given CI, 14.755. Thus, the 95% confidence margin of error is 1.135, so that a 99% confidence margin of error is 2.576(1.135)/1.96 = 1.49. The resulting 99% CI is roughly (13.26,16.25).

(c) The 99% CI in (b) suggests that the average backpack weight as a % of body weight for all 6th graders is between 13.26% and 16.25%. This interval is above the recommendation of (at most) 10%.

15. (a) $\bar{x} \pm 1.96\frac{s}{\sqrt{n}} = 1427 \pm 1.96\frac{325}{\sqrt{514}} = (1398.90, 1455.10)$; we are 95% confident that the average FEV for all such children is between 1398.90 and 1455.10ml.

(b) Given that interval width is 50, the margin of error should be 25. Then,
$n = (1.96^2)(320^2)/25^2 = 629.4 \approx 630.$

17. A large sample two-sided confidence interval for σ would take the form:

$$s \pm (z\ critical\ value)\left(\frac{s}{\sqrt{2n}}\right)$$

A 95% confidence interval for the true standard deviation of the fracture strength distribution is:

$$3.73 \pm (1.96)\left(\frac{3.73}{\sqrt{(2)(169)}}\right)$$

$$= 3.73 \pm .398$$

$$= (3.332, 4.128)$$

19. A 95% upper confidence bound for the true average charge-to-tap time is:

$$\bar{x} + (1.645)\left(\frac{s}{\sqrt{n}}\right)$$

$$382.1 \pm (1.645)\left(\frac{31.5}{\sqrt{36}}\right)$$

$$(382.1 + 8.64) = 390.74\,\text{min}$$

That is, with 95% confidence, the value of μ lies in the interval (0 min, 390.74 min).

21. A 90% lower confidence bound for the true average shear strength is:

$$\bar{x} - (1.28)\left(\frac{s}{\sqrt{n}}\right)$$

$$4.25 - (1.28)\left(\frac{1.30}{\sqrt{78}}\right)$$

$$(4.25 - .188) = 4.062\ kip$$

That is, with 90% confidence, the value of μ lies in the interval (4.062, ∞).

Section 7.3

23. (a) Here, $n = 2343$, and p = 0.53. With 99% confidence, we use the value z* = 2.576 for a two-sided *traditional* confidence interval:

$$\frac{p + \frac{(z^*)^2}{2n} \pm z^* \sqrt{\frac{p(1-p)}{n} + \frac{(z^*)^2}{4n^2}}}{1 + \frac{(z^*)^2}{n}} = \frac{.53 + \frac{2.576^2}{2(2343)} \pm 2.576\sqrt{\frac{.53(1-.53)}{2343} + \frac{2.576^2}{4(2343^2)}}}{1 + \frac{2.576^2}{2343}}$$

$$= \frac{.5314 \pm 2.576(.0266)}{1.003} \Rightarrow (.503, .556)$$

We are 99% confident that between 50.3 and 55.6% of all American adults have watched digitally streamed TV programming on some type of device.

(b) For the width of the interval to be 0.05, we need margin of error to be 0.025. Also, z* for 99% confidence is 2.576.

$$n = \frac{(2.576)^2(.5)(.5)}{.025^2} = 2654.3 \approx 2655$$

25. (a) For a one-sided bound, we need $z_\alpha = z_{.05} = 1.645$; $\hat{p} = \frac{10}{143} = .07$; and $\tilde{p} = \frac{.07 + 1.645^2/(2 \cdot 143)}{1 + 1.645^2/143} = .078$. The resulting 95% lower confidence bound for p, the true proportion of such artificial hip recipients that experience squeaking, is $.078 - \frac{1.645\sqrt{(.07)(.93)/143 + (1.645)^2/(4 \cdot 143^2)}}{1 + (1.645)^2/143} = .078 - .036 = .042$. We are 95% confident that more than 4.2% of all such artificial hip recipients experience squeaking.

(b) If we were to sample repeatedly, the calculation method in (a) is such that p will exceed the calculated lower confidence bound for 95% of all possible random samples of $n = 143$ individuals who received ceramic hips. (We hope that our sample is among that 95%!)

27. (a) Following the same format used for most confidence intervals, i.e., **statistic ± (critical value) (standard error),** an interval estimate for $\pi_1 - \pi_2$ is:

$$(p_1 - p_2) \pm z\sqrt{\frac{p_1(1-p_1)}{n_1} + \frac{p_2(1-p_2)}{n_2}}.$$

(b) The response rate for no-incentive sample: $p_1 = 75/110 = .6818$, while the return rate for the incentive sample is $p_2 = 66/98 = .6735$. Using z = 1.96 (for a confidence level of 95%), a two-sided confidence interval for the true (i.e., population) difference in response rates $\pi_1 - \pi_2$ is:

$$(.6818 - .6735) \pm (1.96)\sqrt{\frac{.6818(1-.6818)}{110} + \frac{(.6735)(1-.6735)}{98}}$$

= .0083 ± .1273 =(-.119, .1356). The fact that this interval contains 0 as a plausible value of $\pi_1 - \pi_2$ means that it is plausible that the two proportions are equal. Therefore, including incentives in the questionnaires does not appear to have a significant effect on the response rate.

(c) Let $\tilde{p}_i$ denote the sample proportion by adding 1 success and 1 failure to the ith sample. We calculate $\tilde{p}_i = (x+1)/(n+2)$, where x is the number of successes (or failures, whichever is desired) in the sample. Then: $\tilde{p}_1 = (75+1)/(110+2) = .67857$ and $\tilde{p}_2 = (66+1)/(98+2) = .67$. Using the format of the equation given in part (a) above, we have the following 95% confidence interval:

$$(\tilde{p}_1 - \tilde{p}_2) \pm z\sqrt{\frac{\tilde{p}_1(1-\tilde{p}_1)}{n_1+2} + \frac{\tilde{p}_2(1-\tilde{p}_2)}{n_2+2}}$$

$$(.67857 - .67) \pm 1.96\sqrt{\frac{.67857(1-.67857)}{110+2} + \frac{.67(1-.67)}{98+2}}$$

$$(.00857) \pm 1.96\sqrt{.004158} \Rightarrow .00857 \pm .1264 \Rightarrow (-.1178, .1350)$$

This interval contains 0, including incentives in the questionnaire does not appear to have a significant effect on the response rate. This is the same conclusion as in part (b).

29. (a) As in Exercise 27, the usual confidence interval format *statistic ± (critical value)(standard error)* gives a confidence interval for:

$\ln(\pi_1/\pi_2)$ of: $\ln(p_1/p_2) \pm z\sqrt{\frac{n_1-u}{n_1 u} + \frac{n_2-v}{n_2 v}}$.

(b) Since we want to estimate the ratio of return rates for incentive group to the non-incentive group, we will call group 1 the incentive group (to match the subscripts in the formula above). The number of returns for the non-incentive group is $v = 75$ out of $n_2 = 110$, so $p_2 = .6818$. For the incentive group, $u = 78$ out of $n_1 = 100$, so $p_1 = .78$. The 95% confidence interval for $\ln(\pi_1/\pi_2)$ is:

$$\ln(.7800/.6818) \pm (1.96)\sqrt{\frac{100-78}{100(78)} + \frac{110-75}{110(75)}} = .1346 \pm .1647$$

= [-.0301, .2993]. To find a 95% interval for π_1/π_2, we exponentiate both end points of this interval, $[e^{-.0301}, e^{.2993}]$ = [.9702, 1.3489], or, about [.970, 1.349]. Since this interval includes the 1, it is plausible that $\pi_1/\pi_2 = 1$; i.e., it is plausible that the return rates are equal. As in Exercise 27, the use of the incentive does not appear to have an effect on the questionnaire return rate.

31. For 90% confidence, the associated z value is 1.645. Since nothing is known about the likely values of π we use .25, the largest possible value of $\pi(1-\pi)$, in the sample size formula: $n = (.25)\left(\frac{1.645}{.05}\right)^2 = 270.6$. To be conservative, we round this value up to the next highest integer and use $n = 271$.

33. Let μ_1 denote the average density of the population having a low percentage of juvenile wood and let μ_2 denote the average density for the population with a moderate percentage of juvenile wood. To estimate the difference between the two means with, say, 95% confidence, we use the large-sample formula: $(\bar{x}_1 - \bar{x}_2) \pm z\sqrt{\frac{s_1^2}{n_1} + \frac{s_2^2}{n_2}} = (.523 - .489) \pm (1.96)\sqrt{\frac{(.0543)^2}{35} + \frac{(.0450)^2}{54}} = .034 \pm .0216 = [.0124, .0556]$. The interval is not too narrow compared to the difference between the average values of the densities, but it does indicate that there seems to be a slight positive difference between the population means (since both endpoints of the interval are positive, indicating that $\mu_1 - \mu_2$ is most likely positive). That is, we can

conclude that the average density for the 'low %' population is probably larger than the average density for the 'moderate %' population.

35. Let μ_1 denote the average toughness for the high-purity steel and let μ_2 denote the average toughness for the commercial purity steel. Then, a lower 95% confidence bound for μ_1-μ_2 is given by: $(\bar{x}_1 - \bar{x}_2) - z\sqrt{\frac{s_1^2}{n_1}+\frac{s_2^2}{n_2}} = (65.6-59.2) - (1.645)\sqrt{\frac{(1.4)^2}{32}+\frac{(1.1)^2}{32}} = 6.4 - .518 = 5.882$. Because this lower interval bound exceeds 5, it gives a reliable indication that the difference between the population toughness levels does exceed 5.

Section 7.4

37. (a) From Table IV (row 10, central area = .95), the critical value is 2.228.

(b) From Table IV (row 20, central area = .95), the critical value is 2.086.

(c) From Table IV (row 20, central area = .99), the critical value is 2.845.

(d) A table entry for df = 50 is not given in Table IV, so we interpolate between the table values for df =40 and df = 60: the value is approximately (2.704 + 2.660)/2 = 2.682

(e) An upper tail area of .01 corresponds to a cumulative area of .99, so from Table IV (row 25, cumulative area area = .99), the critical value is 2.485.

(f) A lower tail area of .025 corresponds to a central area of .95, so from Table IV (row 5, central area = .95), the critical value is 2.571.

39. In the following answers, the critical values are found by selecting the column in Table IV corresponding to the desired cumulative area (which equals the confidence level):

(a) Cumulative area = .95, df = 10: t critical value = 1.812.

(b) Cumulative area = .95, df = 15: t critical value = 1.753.

(c) Cumulative area = .99, df = 15: t critical value = 2.602.

(d) Cumulative area = .99, df = n-1 = 5-1 =4: t critical value = 3.747.

(e) There is no 98% cumulative area column, so we interpolate between the values in the nearby 97.5% and 99% columns: for 24 df, these values are 2.064 and 2.492. The 98% value is *about* 2.064 + (.98-.975)/(.99-.975)[2.492-2.064] = 2.064 + (.3333)[.428] = 2.207.

(f) Since there is no entry for df = n-1 = 38-1 = 37, we must interpolate between the entries for df = 30 and df= 40: critical value ≈ 2.457 + (37-30)/(40-30)[2.423 - 2.457] = 2.4332 or, about 2.43.

41. (a) Hard to tell with such a small sample, but a normal probability plot shows some reason for concern.

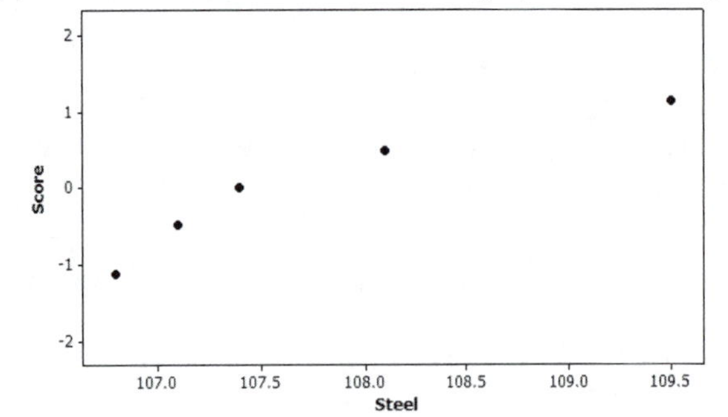

(b) The 95% confidence interval is:

$$107.78 \pm (2.776)\left(\frac{1.0756}{\sqrt{5}}\right)$$

$$107.78 \pm 1.34$$

$$(106.44, 109.12)$$

From this interval 107 does appear to be a plausible value for the true average work of adhesion because it lies in the interval, but 110 does not.

43. (a) The 95% confidence interval is:

$$66221.1 \pm (2.131)\left(\frac{37683.17}{\sqrt{16}}\right)$$

$$(46140,\ 86300)$$

We are 95% confident that the true average mileage of Porshe Boxster cars is between \$46,140 and \$86,300.

(b) The 95% prediction interval is:

$$66221.1 \pm (2.131)\left(37683.17\sqrt{1+\frac{1}{16}}\right)$$

$$(-16554, 148994) \approx (0, 148994)$$

We are 95% confident that the true mileage of any Porshe Boxster cars is between \$0 and \$148,994.

45. (a) The distribution of maximum concrete pressure appears to be somewhat normally distributed.

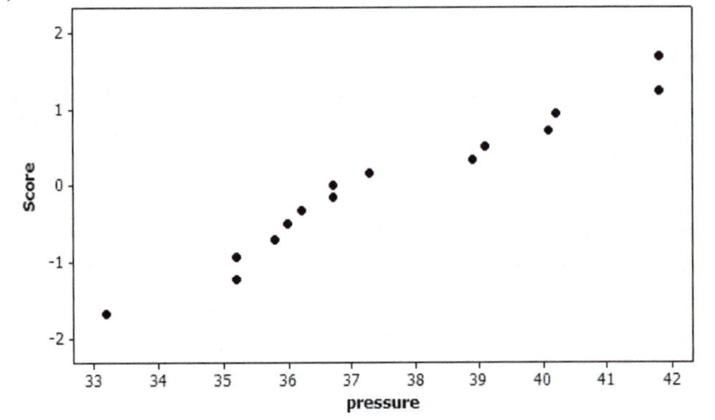

(b) The 95% confidence interval is given by :

$$37.613 \pm (2.145)\left(\frac{2.572}{\sqrt{15}}\right) = (36.19, 39.04)$$ which is same as the SAS interval.

(c) The 95% upper confidence bound is given by :

$$37.613 + (1.761)\left(\frac{2.572}{\sqrt{15}}\right) = 38.78$$

(d) The 95% upper prediction bound is given by :

$$37.613 + (1.761)\left(2.572\sqrt{1+\frac{1}{15}}\right) = 42.29$$

The 95% upper prediction bound is much higher than the 95% confidence bound.

47. Given: $n = 25$
$\bar{x} = .0635$
$s = .0065$

(a) A 95% prediction interval for the amount of warpage of a single piece of laminate is:

$$\bar{x} \pm (\text{t critical value})\left(s\sqrt{1+\frac{1}{n}}\right)$$

The t critical value is obtained from Table IV with df = (n – 1) = 24 for a two-sided interval.

$$.0635 \pm (2.064)\left(.0065\sqrt{1+\frac{1}{25}}\right)$$

$$.0635 \pm .0137$$

(b) For a confidence level of 95%, a two-sided tolerance interval for capturing at least 95% of the warpage values for pieces of laminate in the population sampled is:

$$\bar{x} \pm (\text{tolerance critcal value})s$$

The tolerance critical value is obtained from Table V resulting in:

$$.0635 \pm (2.631)(.0065)$$

$$.0635 \pm .0171$$

$$(.0464, .0806)$$

We can be highly confident that at least 95% of all pieces of laminate have warpage values between .0464 and .0806.

Section 7.5

49. Given:

Type of apples	n	$\overline{x}$	s
Zero-day apples	20	8.74	.66
20-day apples	20	4.96	.39

A 95% confidence interval for the difference between the true firmness of zero-day apples and the true firmness of 20-day apples is:

$$(8.74-4.96)\pm(\text{t critical value})\sqrt{\frac{(.66)^2}{20}+\frac{(39)^2}{20}}$$

The t critical value requires the df be calculated:

$$df=\left[\frac{\left(\frac{.4356}{20}+\frac{.1521}{20}\right)^2}{\frac{\left(.4356/20\right)^2}{19}+\frac{\left(.1521/20\right)^2}{19}}\right]=30.83$$

Thus, we use df = 30 and the t critical value from Table IV is 2.042.

$$3.78\pm(2.042)(.17142)$$
$$3.78\pm.35$$
$$(3.43, 4.13)$$

Thus, with 95% confidence, we can say that the true average firmness for zero-day apples exceeds that of 20-day apples by between 3.43 and 4.13 N.

51. (a)

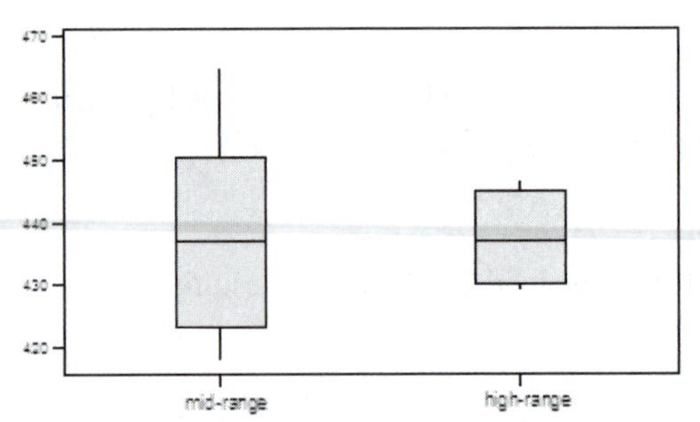

The most notable feature of these boxplots is the larger amount of variation present in the mid-range data as compared to the high-range data. Otherwise, both boxplots look reasonably symmetric and there are no outliers present.

(b) Minitab output:

	n	sample mean	sample standard deviation
Mid-range	17	438.3	15.1
High-range	11	437.45	6.83

A 95% Confidence Interval for (μ mid range - μ high range) is (-7.9 , 9.6)

(Note: df = 23.)

The above analysis was performed by Minitab. The confidence interval was computed as follows:

$$(438.3 - 437.45) \pm (2.069)\sqrt{\frac{(15.1)^2}{17} + \frac{(6.83)^2}{11}}$$

using df = 23, resulting in:

$$.85 \pm 8.69$$
$$(-7.84, 9.54)$$

Since plausible values for $(\mu_1 - \mu_2)$ are both positive and negative (i.e., the interval spans zero) we would conclude that there is not sufficient evidence to suggest that μ_1 and μ_2 differ.

53 The confidence interval can be computed as follows:

$$(19.7 - 10.9) \pm (4.303)\sqrt{\frac{(1.1)^2}{3} + \frac{(0.60)^2}{3}} = (6.498, 11.102) \text{ using df} = 3$$

Because the 95% CI of difference in means does not contain 0, and is positive, we are 95% confident that the true average arsenic level reading using method 1 (new and quick) is higher that the true average arsenic level reading using method 2 (inexpensive field) by between 6.498 and 11.102 μg/L.

55. (a) This exercise calls for a paired analysis. First, compute the difference between number of crashes before and after putting up service signs. These 6 differences are summarized as follows:

$n = 6 \quad \bar{d} = 5.83 \quad s_d = 19.69$, where d = (before value – after value)

Then the t critical value based on df = (n – 1) = 5 must be determined. Table IV provides the t critical value = 2.571.

Thus, a 95% confidence interval for the population mean difference between number of crashes before and after putting up service signs is:

$$6 \pm (2.571)\left(\frac{19.69}{\sqrt{6}}\right) = (-14.83, 26.50)$$

We are 95% confident, that the true average difference in number of crashes before and after putting up service signs is between -14.83 and 26.50. Because the interval contains a 0, it is plausible that the average difference is 0.

(b) We can find a 95% prediction interval: $6 \pm (2.571)\left(19.69\sqrt{1 + \frac{1}{6}}\right) = (-48.68, 60.68)$

We are 95% confident, that the true difference in number of crashes before and after putting up service signs at any site is between -48.68 and 60.68.

57. (a) For the paired data on pitchers, $n = 17$, $\bar{d} = 4.066$, and $s_d = 3.955$. $t_{.025,16} = 2.120$, and the resulting 95% CI is (2.03, 6.10). We are 95% confident that the true mean difference between dominant and nondominant arm translation for pitchers is between 2.03 and 6.10.

(b) For the paired data on position players, $n = 19$, $\bar{d} = 0.233$, and $s_d = 1.603$. $t_{.025,18} = 2.101$, and the resulting 95% CI is (–0.54, 1.01). We are 95% confident that the true mean difference between dominant and nondominant arm translation for position players is between -0.54 and 1.01.

(c) Let μ_1 and μ_2 represent the true mean differences in side-to-side AP translation for pitchers and position players, respectively. We wish to test the hypotheses H_0: $\mu_1 - \mu_2 = 0$ v. H_a: $\mu_1 - \mu_2 > 0$. The data for this analysis are precisely the differences utilized in parts **a** and **b**. Hence, the test statistic is $t = \dfrac{4.066 - 0.233}{\sqrt{\dfrac{3.955^2}{17} + \dfrac{1.603^2}{19}}} = 3.73$. The estimated df = 20 (using software), and the corresponding P-value is P($t >$ 3.73) = .001. Hence, even at the 1% level, we concur with the authors' assessment that this difference is greater, on average, in pitchers than in position players.

59. (a) Let μ_D denote the true mean change in total cholesterol under the aripiprazole regimen. A 95% CI for μ_D, using the "large-sample" method, is 3.75 ± 1.96 (3.878) = (-3.85, 11.35).

(b) Using the "large-sample" procedure again, the 95% CI is $\bar{d} \pm 1.96\frac{s_D}{\sqrt{n}} = \bar{d} \pm 1.96SE(\bar{d})$. If this equals (7.38, 9.69), then midpoint = $\bar{d}$ = 8.535 and width = 2(1.96 $SE(\bar{d})$) = 9.69 – 7.38 = 2.31 ⇒ $SE(\bar{d}) = \dfrac{2.31}{2(1.96)} = .59$. Now, use these values to construct a 99% CI (again, using a "large-sample" z method): $\bar{d} \pm 2.576SE(\bar{d})$ = 8.535 ± 2.576(.59) = 8.535 ± 1.52 = (7.02, 10.06).

Section 7.6

61. (a) To generate a bootstrap interval using a Minitab macro, use the following types of Minitab commands in the macro (for data stored in column c1; set a counter k1 = 1 to start):

```
MTB> sample 20 c1 c10; ← (draws a random sample of size 20)
MTB> replace.    ← (this assures sampling is done with replacement)
MTB> let c2(k1) = mean(c10) ← (store sample mean in column c2)
MTB> let k1 = k1 +1 ← (advance counter by 1)
```

Run your macro any number of times, say 200 or more and then use the stored results in column c2 to form the bootstrap interval. Our interval, based on 200 runs of the above statements gave a 95% bootstrap interval of [.8855, .9590]. Your results may be slightly different since different random samples will be used each time a macro is run. *Note: the answer in the text mistakenly gives a bootstrap interval for the data in Exercise 39 of Chapter 7.*

(b) The t confidence interval in Exercise 43 is [.888, .963] is very close to the bootstrap interval [.886, .959] found in part (a).

63. (a) As we showed in Example 7.15, the sample proportion p = x/n is a maximum likelihood estimator of the population proportion π. That is, the MLE for π is just x/n.

(b) In Section 5.5, we showed that the mean of the sampling distribution of p coincides with π: i.e., $\mu_p = \pi$. Therefore, x/n (the sample proportion) is an unbiased estimator of π.

(c) As shown in part (b) above, x/n is the MLE of π. Using the function $g(\theta) = (1-\theta)^5$ and the Invariance Property of MLEs: since x/n is the MLE for π, then g(x/n) is the MLE for $g(\pi)$. That is, $(1\text{-}x/n)^5$ is the MLE for $(1-\pi)^5$.

65. (a) Denote the 95th percentile of a normal distribution by $x_{.95}$. By definition, then, the area to the left of $x_{.95}$ under the normal curve must be .95; i.e., $P(x \le x_{.95}) = .95$. For a standard normal distribution, the z value with a cumulative area of .95 is approximately 1.645, which is 1.645 standard deviations (since $\sigma = 1$ for the z distribution) to the right of the mean (which is 0 for the z distribution). Therefore, for any normal distribution the value of $x_{.95}$ is 1.645 standard deviations to the right of its mean, so $x_{.95} = \mu + 1.645\sigma$. Next, we know from Example 7.17 that $\bar{x}$ and s^* are the MLEs for μ and σ for a normal distribution (*here* $s^* = s\sqrt{n-1/n}$, where s is the sample standard deviation of the data). Using the function $g(\theta_1, \theta_2) = \theta_1 + 1.645\theta_2$ and the Invariance Property of MLEs, we can then conclude that $g(\bar{x}, s^*) = \bar{x} + 1.645s^*$ is the MLE for $x_{.95}$.

(b) $x_{.95} \approx \bar{x} + 1.645s^* = 384.4 + (1.645)(19.879)\sqrt{10-1/10} = 384.4 + 31.03 = 415.4$.

67. (a) The likelihood function is $L(\lambda,\theta) = \lambda^n e^{-\lambda\sum_1^n (x_i-\theta)}$. After a little simplification, we find $\frac{\partial L}{\partial \theta} = n\lambda^{n+1} e^{-\lambda\sum_1^n (x_i-\theta)}$. Note that this expression is always positive (for $\lambda > 0$), so there is no value of θ that will satisfy $\frac{\partial L}{\partial \theta} = 0$. Instead, we have to take another approach to maximizing $L(\lambda,\theta)$. Taking the partial derivative $\frac{\partial L}{\partial \lambda}$, after much simplification we find: $\frac{\partial L}{\partial \lambda} = \lambda^{n-1} e^{-\lambda\sum_1^n (x_i-\theta)} [n-\lambda\sum_i^n (x_i-\theta)]$, which, when set equal to 0, yields the solution $\lambda = \frac{1}{\bar{x}-\theta}$.

(Note: only the part in the brackets can equal 0, so, we just solve $n-\lambda\sum_i^n (x_i-\theta) = 0$ *for* λ.*)*

Rewriting $L(\lambda,\theta)$ as $\lambda^n e^{-\lambda n\bar{x}} e^{\lambda n\theta}$, we notice that θ only appears in the expression $e^{\lambda n\theta}$, so the value of θ that maximizes $L(\lambda,\theta)$ will have to be one that maximizes $e^{\lambda n\theta}$. Since all data points must exceed θ, then $\theta \le \min(x_1, x_2, ..., x_n)$, so the largest possible value of θ that is allowed by the data is $\theta = \min(x_1, x_2, ..., x_n)$. Notice that this value of θ does not depend on the particular value of λ, so we can conclude that $\hat{\theta} = \min(x_1, x_2, ..., x_n)$ is the MLE of θ. Substituting in to the expression for λ, the MLE of λ is $\hat{\lambda} = \frac{1}{\bar{x}-\hat{\theta}}$.

(b) $\hat{\theta} = \min(x_1, x_2, ..., x_n) = 0.64$ and $\hat{\lambda} = \frac{1}{\bar{x}-\hat{\theta}} = \frac{1}{5.58-.64} = 0.202$.

69. The value $\lambda = 2$ is too large. It essentially puts normal curves with standard deviations of 2s around each data point. That is, normal curves with standard deviations that are twice the standard deviation of the data points themselves. The resulting kernel density will not show much detail in the data.

71. (a) The standard deviation used for the normal curves around each data point will equal $\lambda s = (d/3s)s = d/3$, where d is the minimum distance between any two of the data points. Therefore, the 3-sigma values for these normal curves will equal $\pm 3\sigma = \pm 3(d/3) = \pm d$, so the curves will be so closely packed around each data point that their 3σ values will usually not even overlap. This will lead to a density curve with a very choppy appearance; just a bunch of very small non-overlapping bell curves sitting over the data points.

(b) Larger values of λ will result in smoother kernel density curves.

73. λ will have to be raised. The reason is that the new sample standard deviation, s_{new}, will be slightly smaller than the original standard deviation s_{orig} since the data point removed was an outlier. That is, $s_{orig}/s_{new} > 1$. Therefore, to make the standard deviations used to create the kernel densities approximately equal, $\lambda_{orig}s_{orig} \approx \lambda_{new}s_{new}$, we must have$\lambda_{new}/\lambda_{orig} \approx s_{orig}/s_{new} > 1$.

Supplementary Exercises

75. The following descriptive statistics summarize the tensile strength data.

Variable	N	Mean	Median	TrMean	StDev	SE Mean
Tensile	153	135.39	135.40	135.41	4.59	0.37

Variable	Minimum	Maximum	Q1	Q3
Tensile	122.20	147.70	132.95	138.25

(a) Since the sample size of $n = 153$ is large, we can use the z distribution in calculating the lower confidence bound without making any assumptions about the population distribution. Now, a 95% lower bound is given by: $\bar{x} - z(s/\sqrt{n}) = 135.39 - 1.645(4.59/\sqrt{153}) = 135.39 - 1.645(.371) = 134.78$. We are 95% confidence that the population mean strength is at least 134.78 ksi. (We could have used another confidence level.)

(b) Yes, we must assume that the tensile strength distribution is normal before computing a lower prediction bound. As described at the end of Section 7.4, the normality assumption is required before computing prediction or tolerance intervals.

(c) The tensile strength quantile plot that follows is fairly linear by inspection. Thus, the normality assumption is plausible.

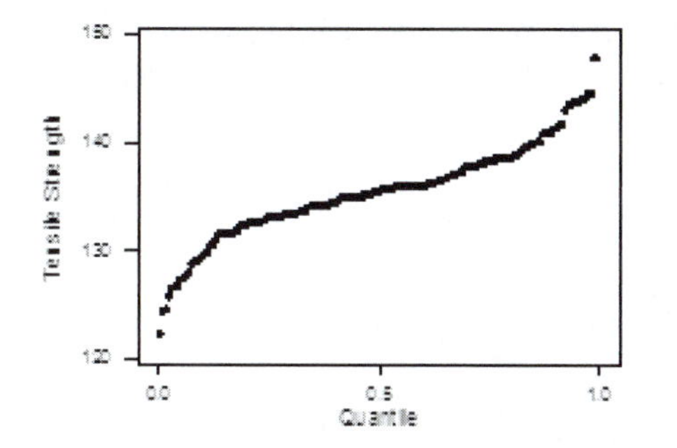

(d) A 95% lower prediction bound is given by $\bar{x} - z \cdot s\sqrt{1+\frac{1}{n}}$. (We can use a z critical value since our sample size is large.) The appropriate z value is then 1.645. So our 95% lower bound is $135.39 - (1.645)(4.59)\sqrt{1+\frac{1}{153}} = 135.39 - 7.635 = 127.81$ We are therefore 95% confident that the next specimen is at least 127.81.

77. The center of any confidence interval for μ_1-μ_2 is always $\bar{x}_1$ - $\bar{x}_2$, so $\bar{x}_1$ - $\bar{x}_2$ = (-473.3 + 1691.9)/2 = 609.3. Furthermore, the half-width of this interval is [1691.9-(-473.3)]/2 = 1082.6. Equating this value to the expression for the half-width of a 95% interval, $1082.6 = (1.96)\sqrt{\frac{s_1^2}{n_1}+\frac{s_2^2}{n_2}}$, we find $\sqrt{\frac{s_1^2}{n_1}+\frac{s_2^2}{n_2}}$ = 1082.6/1.96 = 552.35. For a 90% interval, the associated z value is 1.645, so the 90% confidence interval is then $\bar{x}_1 - \bar{x}_2 \pm (1.645)\sqrt{\frac{s_1^2}{n_1}+\frac{s_2^2}{n_2}}$ = 609.3 ± (1.645)(552.35) = 609.3 ± 908.6 = [-299.3, 1517.9].

79. $n_1 = n_2 = 40$, $\bar{x}_1 = 3975.0$, $s_1 = 245.1$, $\bar{x}_2 = 2795.0$, $s_2 = 293.7$. The large-sample 99% confidence interval for μ_1-μ_2 is then: $\bar{x}_1 - \bar{x}_2 \pm (1.645)\sqrt{\frac{s_1^2}{n_1}+\frac{s_2^2}{n_2}} = (3975.0 - 2795.0) \pm (2.58)\sqrt{\frac{245.1^2}{40}+\frac{293.7^2}{40}} = 1180.0 \pm 156.05 \approx [1024, 1336]$. The value 0 is not contained in this interval so we can state that, with very high confidence, the value of μ_1-μ_2 is not 0, which is equivalent to concluding that the population means are not equal.

81. (a) A normal probability plot of the differences shows no reason for concern. Normality appears to be a reasonable assumption for the distribution of the differences.

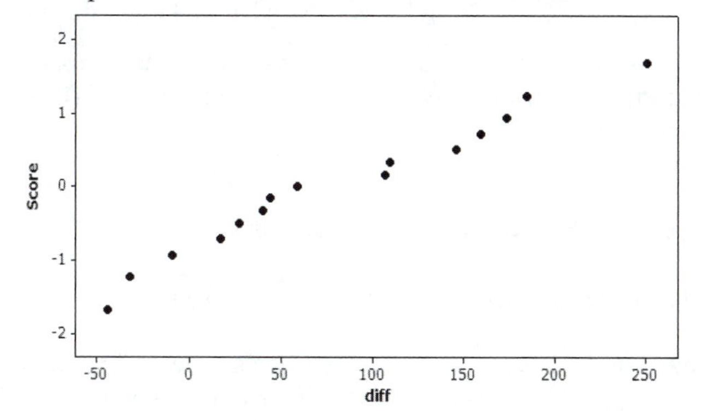

(b) 95% CI (using Paired t in Minitab): (34.1, 130.9). We are 95% confident that on average, the velocity during ER exceeds that during IR by between 34.1 and 130.9deg/sec in all such female collegiate golfers.

Paired T-Test and CI: ER, IR

```
Paired T for ER - IR

N     Mean   StDev   SE Mean
ER            15   -145.3    68.0      17.6
IR            15   -227.8    96.6      24.9
Difference    15     82.5    87.4      22.6

95% CI for mean difference: (34.1, 130.9)
T-Test of mean difference = 0 (vs not = 0): T-Value = 3.66  P-Value = 0.003
```

83. (a) By the definition of the median, $P(x_1 < \tilde{\mu}) = .5$. Of course, $P(x_1 > \tilde{\mu}) = .5$ too, a fact that we use in part (d) below.

(b) P(both observations are smaller than median) = $P(x_1 < \tilde{\mu}$ and $x_2 < \tilde{\mu}) = P(x_1 < \tilde{\mu}) \cdot P(x_2 < \tilde{\mu}) = (.5)(.5) = .25$. Note that we have used the fact that random sampling guarantees that the random variables x_1 and x_2 are independent, which reduces the probability calculation to a simple multiplication.

(c) $y_n = \max(x_1, x_2, x_3, \ldots, x_n)$, so $P(y_n < \tilde{\mu})$ = P(<u>all</u> x_i's are less than $\tilde{\mu}$) = $P(x_1 < \tilde{\mu}) \cdot P(x_2 < \tilde{\mu}) \cdots P(x_n < \tilde{\mu}) = (.5)^n$. As in part (b), independence allows us to multiply the separate probabilities.

(d) $y_1 = \min(x_1, x_2, x_3, \ldots, x_n)$, so $P(y_1 > \tilde{\mu})$ = P(all x_i's are greater then $\tilde{\mu}$)= $P(x_1 > \tilde{\mu}) \cdot P(x_2 > \tilde{\mu}) \cdots P(x_n > \tilde{\mu}) = (.5)^n$.

(e) $P(y_1 < \tilde{\mu} < y_n)$ = 1 - P(either $y_n < \tilde{\mu}$ or $y_1 > \tilde{\mu}$) = $1 - [P(y_n < \tilde{\mu}) + P(y_1 > \tilde{\mu})] = 1 - [(.5)^n + (.5)^n] = 1 - (.5)^{n-1}$. Note that the events $y_n < \tilde{\mu}$ and $y_1 > \tilde{\mu}$ are disjoint, so we <u>can</u> simply use the addition law. Regarding $[y_1, y_n]$ as a confidence interval for $\tilde{\mu}$, we can say that this interval has an associated confidence level of $1 - (.5)^{n-1}$.

(f) For this data, $y_1 = 28.7$ and $y_n = 42.0$, so [28.7, 42.0] is a confidence interval for $\tilde{\mu}$ that has $1-(.5)^{10-1} = .998$, or 99.8 confidence.

(g) Th sample mean and standard deviation of the data are 34.45 and 4.2914, respectively. Based on n-1 = 9 df, the critical t value for a two-sided 99.8% confidence interval is 4.297 (from Table IV), so the desired interval is $\bar{x} \pm t \, s/\sqrt{n} = 34.45 \pm (4.297) \, 4.2914/\sqrt{10}$ $34.45 \pm 5.83 = [28.62, 40.28]$. This interval is a little narrower than the one in (f), which is usually the case as long as the assumption that we are sampling from a normal distribution is valid. Note that the interval in (e) and (f) is valid <u>regardless</u> of the distribution of the population values.

85. (a) x_{n+1} is as likely to be above x_1 as below it, so $P(x_{n+1} > x_1) = .50$.

(b) Any of the three values has an equal chance of being the smallest, so P(x_{n+1} is smallest) = 1/3.

(c) Using the same reasoning as in (b), $P(x_{n+1} < y_1)$ = P(x_{n+1} is the smallest of n random observations) = 1/(n+1). Likewise, $P(x_{n+1} > y_n)$ = P(x_{n+1} is the largest of n random observations) = 1/(n+1).

(d) $P(y_1 < x_{n+1} < y_n) = P(x_{n+1}$ is neither the largest nor smallest value) = 1 - $P(x_{n+1}$ is the smallest or x_{n+1} is the largest) = 1 - [1/(n+1) + 1/(n+1)] = 1 - 2/(n+1). Therefore, the interval $[y_1, y_n]$ is a prediction interval (since it estimates where a single data value is, not where the mean is) with an associated confidence level of 1 - 2/(n+1). For the data of Exercise 77, the interval [28.7, 42.0] would be a 1-2/(n+1) = 1 - 2/(10+1) = .818, or 81.8% prediction interval for the next curing time.

87. The following normal quantile plot shows three points that fall far below a relatively straight line formed by the remaining points. This pattern indicates that population normality is very implausible.

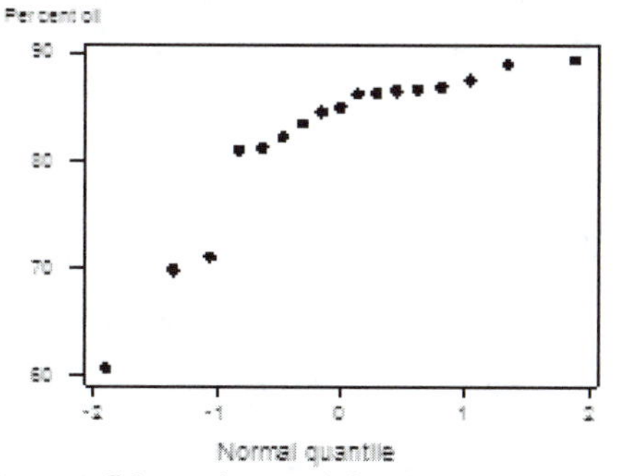

In Exercise 84, a confidence interval for the median was based on the second largest and second smallest observations was shown to have a confidence level of $1-2(n+1)(.5)^n$. For the data of this Exercise, this yields the interval [69.80, 88.80], which has an associated confidence level of $1 - 2(17+1)(.5)^{17} = .9997$, or 99.97%.

89. (a) Using Minitab, we can find the 95% CI to be (38.46, 38.84). We are 95% confident that the true average temperature is between 38.36 and 38.84°C.

One-Sample T: Temp

Variable	N	Mean	StDev	SE Mean	95% CI
Temp	8	38.6500	0.2330	0.0824	(38.4552, 38.8448)

(b) To find a bootstrap sample, we first need to take a large number (say 1000) of samples, each of size 8, with replacement from the original sample. For each of these samples, we compute the sample mean. This will give us 1000 sample means. The 95% bootstrap interval is finding the 2.5th percentile and 97.5th percentile values from these 1000 bootstrap sample means. When we did this for our data, we found the interval to be (38.5, 38.8125) °C.

(Note: The 95% bootstrap interval do not have to be identical. In fact, not often will two people get the exact same interval, because the interval depends of the random samples selected every time from the original sample.)

(c) The 95% t-interval in (a) and the 95% bootstrap interval in (b) – both are fairly close to each other.

91. (a) We shall construct a 95% confidence interval for the true proportion of all American adults who are obese. Here, $n = 4115$, and $p = 1276/4115 \approx .310085$. With 95% confidence, we use the value $z^* = 1.96$ for a two-sided confidence interval:

$$\frac{p + \frac{(z^*)^2}{2n} \pm z^* \sqrt{\frac{p(1-p)}{n} + \frac{(z^*)^2}{4n^2}}}{1 + \frac{(z^*)^2}{n}} = \frac{.310085 + \frac{1.96^2}{2(4115)} \pm 1.96\sqrt{\frac{.310085(1-.310085)}{4115} + \frac{1.96^2}{4(4115^2)}}}{1 + \frac{1.96^2}{4115}}$$

$$\Rightarrow \frac{.31055 \pm 1.96(.007214)}{1.0009} \Rightarrow (.296, .324)$$

We are therefore 95% confident that between 29.6% and 32.4% of all American adults are obese.

(b) Since the interval value dips below 30%, we cannot conclude that the 2002 percentage is more than 1.5 times the 1998 percentage.

Chapter 8

Testing Statistical Hypotheses

Section 8.1

1. (a) Yes, $\sigma > 100$ is a statement about a population standard deviation, i.e., a statement about a population parameter.

   (b) No, this is a statement about the statistic $\bar{x}$, not a statement about a population parameter.

   (c) Yes, this is a statement about the population median $\tilde{\mu}$.

   (d) No, this is a statement about the statistic s (s is the sample, not population, standard deviation.

   (e) Yes, the parameter here is the ratio of two other parameters; i.e, σ_1/σ_2 describes some aspect of the populations being sampled, so it is a parameter, not a statistic.

   (f) No, saying that the difference between two samples means is -5.0 is a statement about sample results, not about population parameters.

   (g) Yes, this is a statement about the parameter λ of an exponential population.

   (h) Yes, this is a statement about the proportion π of successes in a population.

   (i) Yes, this is a legitimate hypothesis because we can make a hypothesis about the population distribution [see (4) at the beginning of this section].

   (j) Yes, this is a legitimate hypothesis. We can make a hypothesis about the population parameters [see (3) at the beginning of this section].

3. Let μ denote the average amperage in the population of all such fuses. Then the two relevant hypotheses are H_0:$\mu = 40$ (the fuses conform to specifications) and H_a:$\mu \neq 40$ (the average amperage either exceeds 40 is less than 40).

5. Let μ denote the average breaking distance for the new system. The relevant hypotheses are H_0:$\mu = 120$ versus H_a:$\mu < 120$, so implicitly H_0 really says that $\mu \geq 120$. A Type I error would be: *concluding that the new system really does reduce the average breaking distance (i.e., rejecting H_0) when, in fact (i.e., when H_0 is true) it doesn't.* A Type II error would be: *concluding that the new system does not achieve a reduction in average breaking distance (i.e., not rejecting H_0) when, in fact (i.e, when H_0 is false) it actually does.*

7. Let μ_1 denote the true average warpage for the regular laminate and let μ_2 denote the true average warpage for the special laminate. The hypotheses of interest are H_0: μ_1-$\mu_2 = 0$ versus H_a: μ_1-$\mu_2 > 0$, so implicitly H_0 asserts that μ_1-$\mu_2 \leq 0$, that is, that the regular laminate does at least as well or better than the special laminate. A Type I error would be: *concluding that the special laminate outperforms the regular laminate (i.e., rejecting H_0) when, in fact (i.e., when H_0 is true) this is not the case.* A Type II error is: *concluding that the special laminate is not better than the regular laminate (i.e., not rejecting H_0) when, in fact (i.e., when H_0 is false) the special laminate really does outperform the regular laminate.*

9. (a) .001 = P-value $\leq \alpha$ = .05, so H_0 should be rejected.
(b) .021 = P-value $\leq \alpha$ = .05, so H_0 should be rejected.
(c) .078 = P-value > α = .05, so H_0 should not be rejected.
(d) .047 = P-value $\leq \alpha$ = .05, so H_0 should be rejected.
(e) .156 = P-value > α = .05, so H_0 should not be rejected.

11. (a) $z = \frac{\bar{x}-5}{s/\sqrt{n}} = \frac{5.23-5}{.89/\sqrt{50}}$ = 1.83, so P-value = area under the z curve to the right (because it's a right-tailed test) of z = 1.83. Therefore, P-value = P(z>1.83) = P(z<-1.83) = .0336.

(b) $z = \frac{\bar{x}-5}{s/\sqrt{n}} = \frac{5.72-5}{1.01/\sqrt{35}}$ = 4.22, so P-value = area under the z curve to the right of z = 4.22. Therefore, P-value = P(z>4.22) ≈ 0.

(c) $z = \frac{\bar{x}-5}{s/\sqrt{n}} = \frac{5.35-5}{1.67/\sqrt{40}}$ = 1.33, so P-value = area under the z curve to the right of z = 1.33. Therefore, P-value = P(z>1.33) = P(z<-1.33) = .0918.

13. (a) This is a test about the population average μ = average silicon content in iron. The null hypothesis value of interest is μ = .85, so the test statistic is of the form $z = \frac{\bar{x}-.85}{s/\sqrt{n}}$. From the wording of the exercise it seems that a 2-sided test is appropriate (since the silicon content is supposed to average .85 and not be substantially larger or smaller than that number), so the relevant hypotheses are $H_0: \mu = .85$ versus $H_a: \mu \neq .85$. We can verify that a 2-sided test was done by calculating the P-value associated with the z value of -.81 given in the printout: the area to the left of -.81 is .2090 (from Table I), so the 2-sided P-value associated with z = -.81 is 2(.2090) = .418 ≈ .42.

(b) The P-value of .42 is quite large, so we don't expect it to lead to rejecting H_0 for any of the usual values of α used in hypothesis testing. Indeed, P = .42 exceeds both α = .05 and α = .10, so in neither case would this data lead to rejecting H_0. It appears to be quite likely that the average silicon content does not differ from .85

15. Let μ denote the true average speedometer reading (at 55 mph). Since we are concerned about whether the speedometer readings may be too high or too low (compared to 55 mph), this requires a 2-sided test of $H_0: \mu = 55$ versus $H_a: \mu \neq 55$. The test statistic would be $z = \frac{\bar{x}-\mu_0}{s/\sqrt{n}} = \frac{53.87-55}{1.36/\sqrt{40}}$ = -5.25. Therefore, P-value = 2P(z < -5.25) ≈ 2(.0000) = 0. Since this P-value is certainly less than α = .01, we reject H_0 and conclude that the average reading is not equal to 55. In particular, the average reading is below 55 so there is a problem with the calibration of the speedometers.

17. (a) Using Minitab, we can find the summary statistics:
Sample mean = 0.7498, sample median = 0.64, and sample standard deviation = 0.3025.

Descriptive Statistics: ALD

Variable	N	N*	Mean	SE Mean	StDev	Minimum	Q1	Median	Q3	Maximum
ALD	49	0	0.7498	0.0432	0.3025	0.3400	0.5150	0.6400	1.0200	1.4400

(b) A normal probability plot of the ALD shows that it is not plausible that the population distribution of ALD is normal. Because the sample size is large enough, normality need not be assumed prior to testing hypotheses about true average ALD.

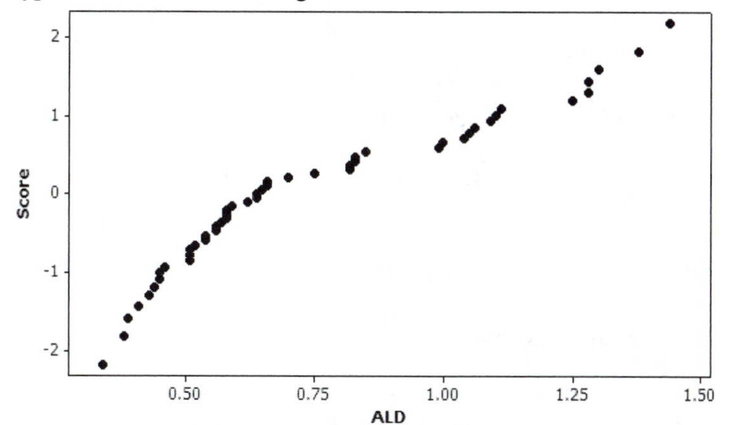

(c) Let μ denote the true average ALD. The relevant hypotheses are:
H_0: $\mu = 1$, versus H_a: $\mu < 1$.

The test statistic is: $z = \left[\dfrac{0.7498-1}{0.3025/\sqrt{49}}\right] = -5.79$. The corresponding P-value = P(z < -5.79) $\approx$ 0.0

Since P-value < α = .05, we reject H_0 and conclude that the true average ALD is less than 1.0. Thus, we conclude that there is strong evidence that true average ALD under such circumstances is less than 1.0.

Section 8.2

19. (a) P-value = P(t > 3.2) = .003 (note: df = (n – 1) = 14)

Since P-value = .003 < α = .05, we reject H_0. We claim $\mu > 20$.

(b) P-value = P(t > 1.8) = .055 (note: df = (n – 1) = 8)

Since P-value = .0055 > α = .01, we fail to reject H_0. We have insufficient evidence to claim $\mu > 20$.

(c) P-value = P(t >-.2) = .578
(note: df = (n – 1) = 23)

Since P-value = .578 >any sensible choice of α, we fail to reject H_0. We have insufficient evidence to claim $\mu > 20$.

21. This situation calls for a two-tailed hypothesis test. The relevant hypotheses are:
H_0: $\mu = .5$ versus H_a: $\mu \neq .5$

(a) P-value = 2P(t > 1.6) = 2(.068) = .136 (note: df = 12)

Since P-value = .136 > α = .05, we fail to reject H_0.

(b) P-value = 2P(t < -1.6) = 2(.068) = .136 (note: df = 12)

Since P-value = .136 > α = .05, we fail to reject H_0.

(c) P-value = 2P(t < -2.6) = 2(.008) = .016 (note: df = 24)

Since P-value = .016 > α = .01, we fail to reject H_0.

(d) P-value = 2P(t < -3.9) $\approx$ 2(0) $\approx$ 0 (note: df = 24)

Since P-value ≈ 0 < even a very small α, we reject H_0.

23. Let μ denote the true average daily energy demand (kW h). The relevant hypotheses are:
H_0: $\mu = 30$, versus H_a: $\mu > 30$.

The test statistic is: $t = \left[\dfrac{32.59 - 30}{10.66/\sqrt{12}}\right] = 0.84$ with d.f. = 12-1 = 11.

From Table VI, the corresponding P-value = P(t > 0.84) = 0.209.
Since P-value = 0.209 > α = .05, we do not reject H_0 and conclude that there is no evidence to contradict the prior belief that true average daily energy demand is at most 30 kW h.

25. Let μ denote the true average melting point (°C). The relevant hypotheses are:
H_0: $\mu = 181$ versus H_a: $\mu > 181$.

Using the sample of $n = 12$ observations, we can calculate the sample mean, $\bar{x} = 181.408$ and sample standard deviation, $s = 0.724$.

The test statistic is: $t = \left[\dfrac{181.408 - 181}{0.724/\sqrt{12}}\right] = 1.95$ with d.f. = 12-1 = 11.

From Table VI, the corresponding P-value = P(t > 1.95) = 0.038.
Since P-value = 0.038 < α = .05, we reject H_0 and conclude that there is convincing evidence that true average melting point is over 181 °C.

27. (a) $\text{df} \approx \dfrac{\left[\frac{5.0^2}{10}+\frac{6.0^2}{10}\right]^2}{\left[\frac{5.0^2}{10}\right]^2\Big/9+\left[\frac{6.0^2}{10}\right]^2\Big/9}$ = 17.43, so round <u>down</u> to df ≈ 17.

(b) $\text{df} \approx \dfrac{\left[\frac{5.0^2}{10}+\frac{6.0^2}{15}\right]^2}{\left[\frac{5.0^2}{10}\right]^2\Big/9+\left[\frac{6.0^2}{15}\right]^2\Big/14}$ = 21.71, so round <u>down</u> to df ≈ 21.

(c) $\text{df} \approx \dfrac{\left[\frac{2.0^2}{10}+\frac{6.0^2}{15}\right]^2}{\left[\frac{2.0^2}{10}\right]^2\Big/9+\left[\frac{6.0^2}{15}\right]^2\Big/14}$ = 18.27, so round <u>down</u> to df ≈ 18.

(d) $\text{df} \approx \dfrac{\left[\frac{5.0^2}{12}+\frac{6.0^2}{24}\right]^2}{\left[\frac{5.0^2}{12}\right]^2\Big/11+\left[\frac{6.0^2}{24}\right]^2\Big/23}$ = 26.08, so round <u>down</u> to df ≈ 26.

29. Let μ_1 denote the true average gap detection threshold for normal subjects and let μ_2 denote the true average gap detection threshold for CTS subjects. Since we are interested in whether the gap detection threshold for CTS subjects exceeds that for normal subjects, a lower tailed test is appropriate. So, we test:

$$H_0 : (\mu_1 - \mu_2) = 0 \quad versus \quad H_a : (\mu_1 - \mu_2) < 0$$

Using the sample statistics provided, the test statistic is:

$$t = \left[\frac{1.71 - 2.53}{\sqrt{\frac{(.53)^2}{8} + \frac{(.87)^2}{10}}}\right]$$

$$t = -2.46 \approx -2.5$$

Using the equation for df provided in the section, the approximate df = 15.1, which we round down to 15.

The corresponding P-value = P(t < -2.5) = .012.

Since the P-value = .012 > α = .01, we fail to reject H_0. We have insufficient evidence to claim that the true average gap detection threshold for CTS subjects exceeds that for normal subjects.

31. (a)

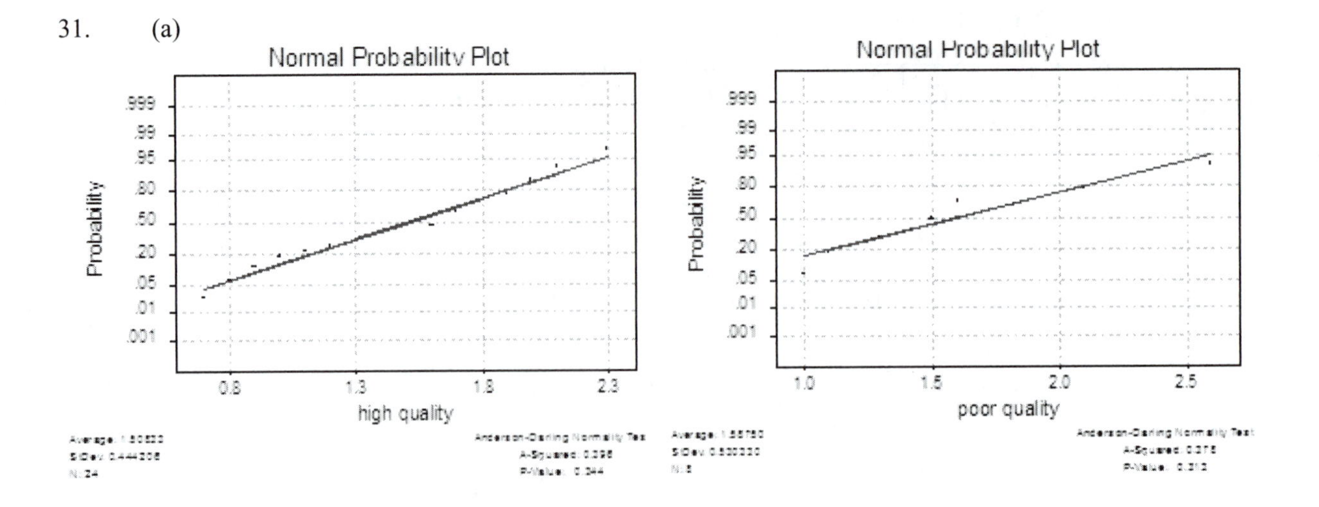

Using Minitab to generate normal probability plots, we see that both plots illustrate sufficient linearity. Therefore, it is plausible that both samples have been selected from normal population distributions.

(b)

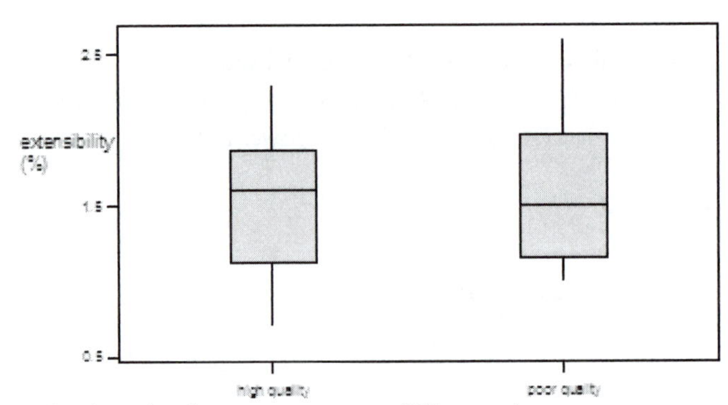

The comparative boxplot does not suggest a difference between average extensibility for the two types of fabrics.

(c) Minitab produced the following analysis:

```
Two sample T for high quality vs poor quality
                N       Mean      StDev   SE Mean
high quality   24      1.508      0.444     0.091
poor quality    8      1.587      0.530      0.19

95% CI for mu high quality - mu poor quality: ( -0.543,  0.38)
T-Test mu high quality = mu poor quality (vs not =): T = -0.38   P-value = 0.71
DF = 10
```

The hypotheses tested were: $H_0 : (\mu_H - \mu_P) = 0 \quad versus\, H_a : (\mu_H - \mu_P) \neq 0$

A two-tailed test was used because we are interested in whether a "difference" exists between the two types of fabrics.

Using the Minitab output, the test statistic, t = -.38. The corresponding df is 10 and the P-value = 2(P(t < -.38)) = .71.

Since the P-value = .71 is extremely large, we fail to reject H_0. That is, there is insufficient evidence to claim that the true average extensibility differs for the two types of fabrics.

33. Let μ_1 denote the true average potential drop for alloy connections and let μ_2 denote the true average potential drop for EC connections. Since we are interested in whether the potential drop is higher for alloy connections, an upper-tailed test is appropriate. So, we test:

$$H_0 : (\mu_1 - \mu_2) = 0 \quad versus\ H_a : (\mu_1 - \mu_2) > 0$$

Using the SAS output provided, the test statistic, when assuming unequal variances, it t = 3.6362, the corresponding df is 37.5, and, the P-value for our upper-tailed test would be [(.0008)/2] = .0004. (Note: P-value = P(t > 3.6362) = .0004.)

Since the P-value = .0004 $< \alpha = .01$, we reject H_0. We have sufficient evidence to claim that the true average potential drop for alloy connections is higher than that for EC connections.

It is possible that, in reaching our conclusion, we have committed a Type I error. That is, we have concluded $\mu_1 > \mu_2$, when in fact, there may be no difference between the two.

35. Let μ_1 denote the true average HAZ depth for high current condition, and let μ_2 denote the true average HAZ depth for non-high current condition. We test:

$$H_0 : (\mu_1 - \mu_2) = 0 \quad versus\ H_a : (\mu_1 - \mu_2) > 0$$

Two-Sample T-Test and CI: High, NonHigh

```
Two-sample T for High vs NonHigh

          N   Mean  StDev  SE Mean
High      9  2.378  0.507     0.17
NonHigh  18  1.926  0.569     0.13


Difference = mu (High) - mu (NonHigh)
Estimate for difference:  0.452
95% lower bound for difference:  0.076
T-Test of difference = 0 (vs >): T-Value = 2.09  P-Value = 0.026  DF = 17
```

Using 2-sample t (assuming samples were drawn from normal distributions) in Minitab (output above), we find the t-test statistic = 2.09, and P-value = 0.026. This P-value > 0.01, so we do not reject H_0. Hence we do not have evidence at a 0.01 significance level that the true average HAZ depth is larger for the high current condition than the non-high current condition.

37. (a) There are two changes that must be made to the procedure we currently use. First, the equation used to compute the value of the t-test statistic is:

$$t = \left[\frac{(\bar{x}_1 - \bar{x}_2) - \Delta}{s_p \sqrt{\frac{1}{n_1} + \frac{1}{n_2}}} \right]$$

where sp^2 is defined in Exercise 51 in Chapter 7.

Secondly, the degrees of freedom = df = $(n_1 + n_2 - 2)$.

(b) Exercise 30, assuming equal variances, happens to produce the same value for the test statistic, t = 3.6362. In general, this does not have to happen. It has happened here because the assumption that $\sigma_1^2 = \sigma_2^2$ is reasonable.
The df = $(n_1 + n_2 - 2) = (20 + 20 - 2) = 38$
The P-value = P(t > 3.6362) with 38 degrees of freedom is (.0008/2) = .0004. This is the same P-value we obtained in part (a). Thus, the conclusion we reach in part (b) is the same as that reached in part (a).

(c) For Exercise 31, the value of the test statistic, t is:

$$t = \left[\frac{(32.8 - 40.5) - (-5)}{s_p\sqrt{\frac{1}{8} + \frac{1}{10}}}\right] = -2.24 \approx -2.2$$

$$where\ s_p = \sqrt{\left(\frac{7}{16}\right)(2.6)^2 + \left(\frac{9}{16}\right)(2.5)^2} = 2.544$$

The df = $(n_1 + n_2 - 2) = (8 + 10 - 2) = 16$. The P-value is P(t < -2.2) = .021.
Since P-value = .021 $> \alpha = .01$, we fail to reject H_0. This is the same conclusion reached in Exercise 31.

39. (a) Let μ_1 be the true average difference in MSD scores (after – before). Then, we test
$H_0: \mu_1 = 0$ versus $H_a: \mu_1 \neq 0$

One-Sample T

```
Test of mu = 0 vs not = 0


 N   Mean   StDev  SE Mean        95% CI            T      P
21  0.190   1.030    0.225  (-0.279, 0.659)  0.85  0.408
```

Using 1-sample t (assuming sample was drawn from normal distribution) in Minitab (output above), we find the t-test statistic = 0.85, and P-value = 0.408, which is > 0.05. Hence we do not have evidence that the true average difference in MSD scores is different from zero.

(b) Let μ_2 be the true average difference in RULA scores (after – before). Then, we test
$H_0: \mu_2 = 0$ versus $H_a: \mu_2 \neq 0$

One-Sample T

```
Test of mu = 0 vs not = 0


 N     Mean   StDev  SE Mean         95% CI             T      P
21   -1.520   1.560    0.340  (-2.230, -0.810)   -4.47  0.000
```

Using 1-sample t (assuming sample was drawn from normal distribution) in Minitab (output above), we find the t-test statistic = -4.47, and P-value < 0.05. Hence we have strong evidence that the true average difference in RULA scores is different from zero.

(c) Measurements were taken before and after intervention. The intervention in the form of a short oral presentation would most likely not lead to instant reductions in musculoskeletal disorders (MSD). However, such an intervention could cause an immediate change in one's posture and therefore have a major impact on one's RULA score. Thus, it would be easier to improve RULA score in a short period of time, rather than the MSD score.

41. Each worker used both shovels, so the data is paired.
Let μ_d be the true average amount by which energy expenditure using the conventional shovel exceeds that using the perforated shovel. We want to test H_0: $\mu_d = 0$ versus H_a: $\mu_d > 0$,

The observed differences (Conventional minus Perforated) are: 0.0, 0.0004, -0.0001, 0.0009, -0.0001, -0.0001, 0.0004, 0.0, 0.0002, 0.0001, 0.0006, 0.0002, 0.0007. The sample mean and standard deviation of these observations are $\bar{d}$ = 0.000246 and s_d = 0.000331. The test statistic is:

$$t = \frac{\bar{d} - 0}{s_d/\sqrt{n}} = \frac{0.000246 - 0}{0.000331/\sqrt{13}} = 2.68.$$

For d.f. = n-1 = 13-1 = 12, we can use Table VI to find the P-value for this test: P-value = $P(t > 2.68)$ = 0.01. Because the P-value = 0.01 < α = 0.05, we reject H_0. We conclude that there is evidence that true average energy expenditure using the conventional shovel exceeds that using the perforated shovel.

43. Let μ be the true average % difference of measured from stated energy content of supermarket meals. We want to test H_0: $\mu = 0$ versus H_a: $\mu \neq 0$.

To perform the calculations, first compute the percent changes by using the formula $\frac{measured - stated}{stated} \times 100$. For the sample of 10 observations, the %differences were found to be 17.78, 45, 21.58, 33.04, 5.5, 16.49, 15.2, 10.42, 81.25, 26.67. Thus, n = 10, sample mean, $\bar{x}$ = 27.29, and sample standard deviation, s = 22.12.

Then, the test statistic can be calculated as

$$t = \frac{27.29 - 0}{22.12/\sqrt{10}} = 3.90.$$

For d.f. = n-1 = 10-1 = 9, we can use Table VI to find the P-value for this test: P-value = $2P(t > 3.90)$ = 0.004. Because the P-value = 0.004 < α = 0.05, we reject H_0. We conclude that there is evidence that true average % difference of measured from stated energy content of supermarket meals differs from zero.

Section 8.3

45. Using the number 1 (for business), 2 (for engineering), 3 (for social science), and 4 (for agriculture), let π_i = the true proportion of all clients from discipline i. If the Statistics Department's expectations are correct, then the relevant null hypothesis is:

$H_0: \pi_1 = .40 \quad \pi_2 = .30 \quad \pi_3 = .20 \quad \pi_4 = .10$ *versus*

H_a: The Statistics Department's expectations are not correct

Using the proportions in H_0, the expected number of clients are:

Client's discipline	Expected number of clients
Business	(120)(.40) = 48
Engineering	(120)(.30) = 36
Social science	(120)(.20) = 24
Agriculture	(120)(.10) = 12

Since all expected counts are at least 5, the chi-squared test can be used. The value of the χ^2 test statistic is:

$$\chi^2 = \left[\sum \frac{(\text{observed} - \text{expected})^2}{\text{expected}}\right]$$

$$= \left[\frac{(52-48)^2}{48} + \frac{(38-36)^2}{36} + \frac{(21-24)^2}{24} + \frac{(9-12)^2}{12}\right] = 1.57$$

For df = (k – 1) = (4 – 1) = 3, the corresponding P-value = $P(\chi^2 > 1.57)$. Using Table VII, we find that the P-value > .10. Since the P-value is larger than $\alpha = .05$, we fail to reject H_0. We have no evidence to suggest that the Statistics Department's expectations are incorrect.

47. Using the numbering 1 (for African American), 2 (Asian), 3(Caucasian), and 4 (Hispanic), let π_i = the true proportion in which i^{th} ethnic group is represented in all commercials. We state our hypotheses as: H_0: $\pi_1 = 0.177$, $\pi_2 = 0.032$, $\pi_3 = 0.734$, and $\pi_4 = 0.057$. The alternative hypothesis in this case would be H_a: *At least one of the proportions does not equal the census proportion*. Using the proportions in H_0, the expected numbers of homicides (out of n = 404) are shown below the actual numbers from the problem:

Ethnicity	**AfAm.**	**Asian**	**Caucasian**	**Hispanic**	**total #**
Observed	57	11	330	6	404
Expected	71.51	12.93	296.54	23.03	404

χ^2 contributions are: 2.94, 0.29, 3.78, 12.59

The χ^2 test statistic value is $\chi^2 = 2.94 + 0.29 + 3.78 + 12.59 = 19.6$. For d.f = k-1 = 4-1 = 3, the value 19.6 is larger than any of the entries in Table VII (column df=3), so the P-value < 0.001. Therefore, since P-value < .001 < .01 = α, H_0 is rejected and we conclude that these data do provide evidence that the proportions in commercials are different from the census proportions.

49. Using the numbering 1 through 9, for the nine medal pairs, let π_i = the true probability of ith medal pair. We state our hypotheses as:

H_0: $\pi_1 = \pi_2 = \ldots = \pi_9 = 1/9$. The alternative hypothesis in this case would be H_a: *At least one of the proportions does not equal 1/9.*

Using Chi-square Goodness of Fit test (Univariate Data) in Minitab (output below), we find the χ^2-test statistic = 0.660621, and P-value =0.719 > 0.10. Hence, at a significance level of .10, we have no evidence that wine experts and consumers differ in their ratings more than what we expect by random chance.

Chi-Square Goodness-of-Fit Test for Observed Counts in Variable: obsd

```
                          Test                Contribution
Category   Observed  Proportion  Expected       to Chi-Sq
same             69       0.333    71.928        0.119191
diffbyone       102       0.445    96.120        0.359700
diffbytwo        45       0.222    47.952        0.181730

N   DF    Chi-Sq  P-Value
216   2  0.660621    0.719
```

51. We can arrange the data in a two-way table as follows:

	Counseling	No counseling	Total
Side effects	24	8	32
None	31	44	75
	55	52	107

Null hypothesis, H_0: No association between subgroup type and whether or not develop side effects
Alternative hypothesis, H_a: There is an association between subgroup type and whether or not develop side effects

Using Chi-square test for two-way tables in Minitab (output below), we find the χ^2-test statistic = 10.177, df = (2-1)x(2-1) = 1, and P-value =0.001. Hence, with the small P-value, we have strong evidence that there is an association between subgroup type and whether or not develop side effects.

Chi-Square Test: counsel, no counsel

```
Expected counts are printed below observed counts
Chi-Square contributions are printed below expected counts
                        no
          counsel  counsel  Total
SideEff        24        8     32
            16.45    15.55
            3.467    3.667

None           31       44     75
            38.55    36.45
            1.479    1.564

Total          55       52    107

Chi-Sq = 10.177, DF = 1, P-Value = 0.001
```

53. H_0: No association between gender and neck pain
H_a: There is an association between gender and neck pain

Using Chi-square test for two-way tables in Minitab (output below), we find the χ^2-test statistic = 142.122, df = (3-1)x(2-1) = 2, and P-value < 0.01. Hence, with the small P-value, we have strong evidence that there is an association between gender and neck pain.

Chi-Square Test: men, women

```
Expected counts are printed below observed counts
Chi-Square contributions are printed below expected counts

            men    women  Total
never      3048     1842   4890
        2814.15  2075.85
         19.433   26.345

occ        1767     1411   3178
        1828.91  1349.09
          2.095    2.841

often       590      734   1324
         761.95   562.05
         38.804   52.604

Total      5405     3987   9392

Chi-Sq = 142.122, DF = 2, P-Value = 0.000
```

55. (a) The expected cell count for any cell can be found using the formula = $\frac{\text{(row total)x(column total)}}{\text{total sample size}}$

Expected cell counts table

	TE	TT	Total
Received prescript.	1.56	2.44	4
Didn't	32.44	50.56	83
	34	53	87

Yes, two of the expected counts are smaller than 5. Thus, instead of a Chi-square test for two way tables, we should use Fisher's exact test.

(b) Using Minitab, 2 proportion z-test *but reading off the* P-value *corresponding to Fisher's exact test:*

Test and CI for Two Proportions

```
Sample  X   N  Sample p
1       4  34  0.117647
2       0  53  0.000000

Difference = p (1) - p (2)
Estimate for difference:  0.117647
95% CI for difference:  (0.00934908, 0.225945)
Test for difference = 0 (vs not = 0):  Z = 2.13  P-Value = 0.033

* NOTE * The normal approximation may be inaccurate for small samples.

Fisher's exact test: P-Value = 0.021
```

Fisher's exact test P-value = 0.021.

Hence, at a 5% significance level, we have convincing evidence that surgery method affects the provision of ondansetron prescriptions.

Section 8.4

57. (a) i. Since $r^* > .9347$, P-value $> .10$
ii. Since $.8804 < r^* < .9180$, $.01 <$ P-value $< .05$
iii. Since $r^* > .9662$, P-value $> .10$
iv. Since $r^* < .9408$, P-value $< .01$

(b) i. Fail to reject H_0, since P-value $> .05$
ii. Reject H_0, since P-value $< .05$
iii. Fail to reject H_0, since P-value $> .05$
iv. Reject H_0, since P-value $< .05$

59. The Ryan-Joiner test P-value is larger than .10, so we conclude that this data could reasonably have come from a normal population. We can safely use a one-sample t-test to test hypotheses about the value of the true average compressive strength.

61. (a) The hypotheses are
H_0: The distribution of %IACS is normal, versus, H_a: The distribution of %IACS is not normal

The Ryan-Joiner test P-value < 0.010, which is smaller than a 0.05 significance level; reject H_0. Thus, it is not plausible that the distribution of %IACS is normal.

Mean	33.13
StDev	13.17
N	8
RJ	0.774
P-Value	<0.010

(b) After removing the %IACS value of 65, the Ryan-Joiner test P-value > 0.100, which is larger than a 0.05 significance level; do not reject H_0. Thus, it is plausible that the distribution of %IACS is normal.

Mean	28.57
StDev	2.992
N	7
RJ	0.996
P-Value	>0.100

63. The equation $\mu = (1/\lambda) - \frac{x_0 e^{-\lambda x_0}}{1-e^{-\lambda x_0}}$ can not be explicitly solved for λ. Instead, we replace μ and x_0 by the given values $\bar{x} = 13.086$ and $x_0 = 70$, then solve numerically for the estimated value $\hat{\lambda}$. That is, we solve the equation $13.086 = (1/\hat{\lambda}) - \frac{70e^{-70\hat{\lambda}}}{1-e^{-70\hat{\lambda}}}$ numerically for $\hat{\lambda}$. The solution is $\hat{\lambda} \approx .0742$. With (a_{i-1}, a_i) denoting the ith class interval ($i = 1,2,3,\ldots,9$), the expected class frequencies are given by : (n)(proportion in i^{th} class) $= (40)\int_{a_{i-1}}^{a_i} f(x)dx = 40\left[\frac{e^{-.0742a_{i-1}} - e^{-.0742a_i}}{1-e^{-.0742(70)}}\right]$. The expected frequencies are: 18.0, 9.9, 5.5, 3.0, 1.8, .9, .5, .3, and .1. From these expected values we obtain the calculated χ^2 value of 1.34. Using d.f. $= k - 1 - m = 9 - 1 - 1 = 7$, (m =1 because we estimated the single parameter λ from the data), the value 1.34 is much smaller than any of the entries in the df = 7 column of Table VII, so the P-value must exceed .10.

In fact, the exact P-value is much larger still, so H_0 is not rejected and we conclude that it is reasonable to assume that this data came from a truncated exponential population.

Section 8.5

65. (a)

Sample size, n	z value	P-value (upper)
100	$\frac{(\bar{x}-\mu_0)}{\sigma/\sqrt{n}} = \frac{(101-100)}{15/\sqrt{n}} = \sqrt{n}/15 = \sqrt{100}/15 \approx .67$	.2514
400	$\sqrt{n}/15 = \sqrt{400}/15 \approx 1.33$	.0918
1600	$\sqrt{n}/15 = \sqrt{1600}/15 \approx 2.67$	.0038
2500	$\sqrt{n}/15 = \sqrt{2500}/15 \approx 3.33$	.0004

(b) β = P(Type II error) = P(don't reject H_0 when H_0 is false) = $P(\bar{x} < 100+(2.33)(15)/\sqrt{n})$, when μ = 101) = $P(z < (100+(2.33)(15)/\sqrt{n} - 101)/(15/\sqrt{n})) = P(z < 2.33 - \sqrt{n}/15)$.

Sample size	β
100	$P(z < 2.33 - \sqrt{100}/15) = P(z < 1.66) = .9515$
400	$P(z < 2.33 - \sqrt{400}/15) = P(z < 1.00) = .8413$
1600	$P(z < 2.33 - \sqrt{1600}/15) = P(z < -.33) = .3707$
2500	$P(z < 2.33 - \sqrt{2500}/15) = P(z < -1.00) = .1587$

Because the value μ = 101 is so close to the null hypothesis value μ = 100, all the β values above are fairly large.

67. We are testing the two-sided hypotheses $H_0: \mu = 100$ versus $H_a: \mu \neq 100$ at the $\alpha = .01$ significance level when n=15 is proposed. The standard deviation σ is thought to be between .8 and .1. The first two printouts show that the power of detecting shifts of $.5\sigma$ or $.8\sigma$ will be very low. The second printout shows that power can be increased to 90% by increasing the sample size to 42 for a $.5\sigma$ shift and 19 for a $.8\sigma$ shift.

69. Combining both samples into one group and ranking:

Observation:	179	183	216	229	232	245	247	250	286	299
Sample #:	2	2	2	1	2	1	2	1	1	1
Rank:	1	2	3	4	5	6	7	8	9	10

The Wilcoxon test statistic is w = sum of ranks for first sample = 4+6+8+9+10 = 37. The total possible sum of ranks in a sample of size n_1 = 5 varies between a minimum of 1+2+3+4+5 = 15 and a maximum of 6+7+8+9+10 = 40. Because this is a 2-sided test, we include values of w that are as close to the minimum as the value of 37 is to the maximum possible sum. Therefore, the P-value = $P(w \geq 37) + P(w \leq 18)$ = .028 +.028 = .056. For the hypotheses H_0:μ_1-μ_2 = 0 versus H_a: μ_1-μ_2 = 0, we conclude at significance level, say α = .05, that H_0 should not be rejected. There is not sufficient evidence to conclude that the two average bond strengths are different.

Supplementary Exercises

71. (a) The hypotheses can be stated as:
H_0: The distribution of amount left is normal, versus, H_a: The distribution of amount left is not normal

Using Minitab, a Ryan-Joiner test gives $r^* = 0.985$, and P-value > 0.100. Hence, it is plausible that the data are from a normal distribution.

Mean	0.502
StDev	0.1023
N	5
RJ	0.985
P-Value	>0.100

(b) Let μ = true average amount left. Then we can test the hypotheses: H_0: $\mu = 0.6$ versus H_a: $\mu < 0.6$. Using 1-sample t (because safe to assume sample was drawn from normal distribution) in Minitab (output below), we find the t-test statistic = -2.14, and P-value = 0.049 > 0.01. Hence we do not have evidence that the true average %difference is different from zero.

One-Sample T: amt left

```
Test of mu = 0.6 vs < 0.6

                                                99% Upper
Variable  N    Mean   StDev  SE Mean      Bound       T      P
amt left  5  0.5020  0.1023   0.0458     0.6735   -2.14  0.049
```

(c) Type I error would be concluding that the true average amount left is less than 10% of the advertised contents, when it is not.

Type II error would be concluding that the true average amount left is not different from 10% of the advertised contents, when the true average amount left is less than 10% of the advertised contents.

In this case, a Type II error might have been made.

73. (a) No, it does not appear plausible that the distribution is normal. Notice that the mean value, $\bar{x} = 215$, is not nearly in the middle of the range of values, 5 to 1176. The midrange would be about 585. Since the mean is so much lower than this, one would suspect the distribution is positively skewed.

However, it is not necessary to assume normality if the sample size is "large enough", due to the central limit theorem. Since this problem has a sample size which is "large enough" (i.e., 47 > 30), we can proceed with a test of hypothesis about the true mean consumption.

(b) Let μ denote the true mean consumption. Since we are interested in determining if there is evidence to contradict the prior belief that μ was at most 200 mg, the following hypotheses should be tested.

H_0: $\mu = 200$ versus H_a: $\mu > 200$.
The value of the test statistic is:

$$z = \left(\frac{\bar{x} - 200}{s/\sqrt{n}} \right) = \left(\frac{215 - 200}{235/\sqrt{47}} \right) = .44$$

The corresponding P-value = P(z > .44) = .33.

Since P-value = .33 > most any choice of α, we fail to reject H_0. There is insufficient evidence to suggest that the true mean caffeine consumption of adult women exceeds 200 mg per day.

75. This analysis should be conducted using a t-test, if one can assume that the data, the neutralizing amount of antitoxin, could have come from a normal distribution.

Let μ denote the true mean neutralizing amount of antitoxin. Since we are interested in determining if a previous value for μ can be contradicted, the relevant hypotheses are:

H_0: $\mu = 1.75$ versus H_a: $\mu \neq 1.75$.
The value of the test statistic is:

$$t = \left(\frac{1.89 - 1.75}{.42/\sqrt{26}} \right) = 1.70$$

Since we are conducting a two-tailed test, the corresponding P-value = 2P(t > 1.70) when df = (n – 1) = 25.

Using Table VI, we find that the P-value = 2(.051) = .102.

Since P-value = .102 > α = .05, we fail to reject H_0. We have insufficient evidence to contradict the prior research claim that the true mean neutralizing amount of antitoxin is 1.75.

77. Let π denote the true proportion of front-seat occupants involved in head-on collisions, in a certain region, who sustain no injuries. Given the wording of the exercise, the relevant hypotheses are:

H_0: $\pi = \left(\frac{1}{3}\right)$ versus H_a: $\pi < \left(\frac{1}{3}\right)$

[Note: $(319)\left(\frac{1}{3}\right) = 106.3$ and $(319)\left(\frac{2}{3}\right) = 212.7$ are each ≥ 5].

So, the test statistic is:

$$z = \left[\frac{p - \pi_0}{\sqrt{\frac{(\pi_0)(1-\pi_0)}{n}}} \right] = \left[\frac{\left(\frac{95}{319}\right) - \left(\frac{1}{3}\right)}{\sqrt{\frac{\left(\frac{1}{3}\right)\left(\frac{2}{3}\right)}{319}}} \right] = -1.35$$

The corresponding P-value = P(z < -1.35) = .0885

Since P-value = .0885 > α =.05, we fail to reject H_0. We have insufficient evidence to claim that less than one-third of all such accidents result in no injuries.

79. First, we will assume that the sample size of 30 for each of the two samples is sufficiently large to use the Central Limit Theorem in determining our test statistic. This assumption may not be a sensible assumption if the two underlying distributions are excessively skewed.

Let μ_1 denote the true mean headability rating for aluminum killed steel specimens and μ_2 denote the true mean headability rating for silicon killed steel.

Given that we are interested in assessing if there is a difference between the two steel types, the relevant hypotheses are:

$H_0 : (\mu_1 - \mu_2) = 0 \quad versus \quad H_a : (\mu_1 - \mu_2) \neq 0$

The test statistic is:

$$z = \left[\frac{(\bar{x}_1 - \bar{x}_2)}{\sqrt{\frac{s_1^2}{n_1} + \frac{s_2^2}{n_2}}} \right] = \frac{(7.09 - 6.43)}{\sqrt{\frac{(1.19)^2}{30} + \frac{(1.08)^2}{30}}} = 2.25$$

The corresponding P-value = 2P(z > 2.25) = 2(.0122) = .0244

Since P-value = .0244 < α = .05, we reject H_0. We have sufficient evidence to claim that there is a difference in headability ratings. We agree with the article's authors.

[Note: Had you not wished to use the z test statistic and instead used a two-sample t-test, the df = 57 and the P-value = 2P(t > 2.2) $\approx$ 2(.014) = .028.]

81. (a) Let μ_1 denote the true mean strength for males and μ_2 denote the true mean strength for females. The hypotheses tested here were:

$H_0 : (\mu_1 - \mu_2) = 0 \quad versus \quad H_a : (\mu_1 - \mu_2) \neq 0$

If one assumes equal population variances, and uses the pooled sample variance you will obtain: s_p = 31.77, t = 2.47, and df = 24. The corresponding P-value for this test is: 2P(t > 2.5) = 2(.010) = .02. These values are quite close to those reported in the exercise.

Notice, however, that the assumption of equal population variances and the t-test statistic that accompanies this assumption is not described in the body of the chapter.

If one uses the method described in the body of the chapter, then t = 2.84 and df = 18. This results in a P-value = 2(P(t > 2.8) = 2(.006) = .012.

(b) Revise the hypotheses:

$$H_0 : (\mu_1 - \mu_2) = 25 \quad versus \quad H_a : (\mu_1 - \mu_2) > 25$$

The test statistic (without assuming equal population variances) is:

$$t = \left[\frac{(129.2 - 98.1) - 25}{\sqrt{\frac{(39.1)^2}{15} + \frac{(14.2)^2}{11}}} \right] = .556$$

With df = 18, the P-value = P(t > .6) = .278

Since the P-value is greater than any sensible choice of α, we fail to reject H_0. There is insufficient evidence that the true average strength for males exceeds that for females by more than 25N.

83. (a) Let μ_1 be the true average force in a dry medium at a higher temperature and μ_2 be the true average force in a dry medium at a higher temperature. Then, the hypotheses are:
H_0: $\mu_1 - \mu_2 = 100$ versus H_0: $\mu_1 - \mu_2 > 100$

Using 2-sample t (assuming samples were drawn from normal distributions) in Minitab (output below), we find the t-test statistic = 2.58, and P-value = 0.015, which is < 0.05. Hence we have convincing evidence that the true average force in dry medium at a higher temperature exceeds that at lower temperature by more than 100N.

Two-Sample T-Test and CI

```
                            SE
Sample  N   Mean  StDev   Mean
1       6  325.7   35.0     14
2       6  170.6   39.1     16


Difference = mu (1) - mu (2)
Estimate for difference:  155.1
95% lower bound for difference:  115.9
T-Test of difference = 100 (vs >): T-Value = 2.58  P-Value = 0.015
```

(b) Let μ_1 be the true average force in a wet medium at a lower temperature and μ_2 be the true average force in a wet medium at a higher temperature. Then, the hypotheses are:
H_0: $\mu_1 - \mu_2 = 50$ versus H_0: $\mu_1 - \mu_2 > 50$

Using 2-sample t (assuming samples were drawn from normal distributions) in Minitab (output below), we find the t-test statistic = 0.46, and P-value = 0.328, which is fairly large. Hence we do not have evidence that the true average force in wet medium at a lower temperature exceeds that at higher temperature by more than 50N.

Two-Sample T-Test and CI

```
                            SE
Sample  N   Mean  StDev   Mean
1       6  366.4   34.8     14
2       6  306.1   42.0     17


Difference = mu (1) - mu (2)
Estimate for difference:  60.3
95% lower bound for difference:  19.5
T-Test of difference = 50 (vs >): T-Value = 0.46  P-Value = 0.328
```

85. Since each patient was given both the drug and a placebo, the data is paired. So, a paired t-test should be conducted.

First compute the 14 differences (deanol – placebo). $\bar{d} = .821$ $s_d = 2.52$.

Let μ_d denote the true average difference in the total severity index for deanol versus the placebo. The relevant hypotheses are: $H_0 : \mu_d = 0$ *versus* $H_a : \mu_d > 0$

The value of the test statistic is:

$$t = \left[\frac{(.821-0)}{2.52/\sqrt{14}}\right] = 1.22$$

. With df = (n – 1) = 13, the P-value = P(t > 1.2) = .126.

Since the P-value is larger than any sensible choice of α, we fail to reject H_0.

There is insufficient evidence to claim that, on average, deanol yields a higher total severity index than does the placebo treatment.

87. Let μ_1 denote the true average maximum lean angle for young females, and let μ_2 denote the true average maximum lean angle for old females. We shall test $H_0 : \mu_1 - \mu_2 = 10$ versus $H_a : \mu_1 - \mu_2 > 10$ at the $\alpha = .10$ significance level.

The relevant sample statistics from both samples include the following:

$\bar{x}_1 = 31.7$, $s_1 = 3.057$, $n_1 = 10$, $se_1 = 3.057/\sqrt{10} = .9667$

$\bar{x}_2 = 16.2$, $s_2 = 4.438$, $n_2 = 5$, $se_2 = 4.438/\sqrt{5} = 1.9847$

The test-statistic is given by: $t = \dfrac{(\bar{x}_1 - \bar{x}_2) - \Delta}{\sqrt{\frac{s_1^2}{n_1} + \frac{s_2^2}{n_2}}} = \dfrac{(31.7 - 16.2) - 10}{\sqrt{\frac{3.057^2}{10} + \frac{4.438^2}{5}}} = 2.49$

The degrees of freedom are given by:

$$\text{df} = \frac{[(se_1)^2 + (se_2)^2]^2}{\frac{(se_1)^4}{n_1 - 1} + \frac{(se_2)^4}{n_2 - 2}} = \frac{[.9667^2 + 1.9847^2]^2}{\frac{.9667^4}{10-1} + \frac{1.9847^4}{5-1}} = \frac{23.753}{3.976} = 5.97\text{, so round down to 5 df}$$

At $\alpha = .10$ and 5 degrees of freedom, the corresponding *t* critical value is 1.476. Since our test-statistic of 2.49 is greater than this critical value, we reject H_0. We conclude that the true average maximum lean angle for older females is more than 10 degrees smaller than that of younger females.

89. (a)

Ranking	Observed	Expected	(O-E)²/E
First	22	48(.27477) =13.18896	5.8863
Second	10	48(.20834) = 10.0032	0
Third	5	48(.15429) = 7.40592	.781598
Fourth or Lower	11	48(.3626) = 17.4048	2.357

$$X^2 = \sum_{\text{all categories}} \frac{(\text{observed} - \text{expected})^2}{\text{expected}} = 5.8863 + 0 + .781598 + 2.357 = 9.02$$

Because we have 4 categories, we have 4 – 1 = 3 degrees of freedom. From the chi-square table, 7.81 < 9.02 < 11.34, and so our P-value is between .01 and .05. Thus, the above model is questionable.

(b)

Ranking	Observed	Expected	(O-E)²/E
First	22	48(.45883) = 22.02384	.000026
Second	10	48(.18813) = 9.03024	.104143
Third	5	48(.11032) = 5.29536	.016474
Fourth or Lower	11	48(.24272) = 11.65056	.036327

$$X^2 = \sum_{\text{all categories}} \frac{(\text{observed} - \text{expected})^2}{\text{expected}} = .000026 + .104143 + .0164474 + .036327 + .15697$$

Our X^2 statistic is quite small, and so the proposed model appears to fit the data quite well.

Chapter 9

The Analysis of Variance

Section 9.1

1. (a) H_0: $\mu_A = \mu_B = \mu_C$; where μ_i = average strength of wood of Type i.

   (b) When a null hypothesis is not rejected in an ANOVA test it can often be good news. When there is no significant difference between two populations (types of beams in this case), you are then free to use other factors when making decisions about the populations. For instance, in this exercise, the choice is narrowed to Type A or B (both of which were shown by the ANOVA test to be superior in strength to Type C). To make the final decision (between A and B), we might use another factor, such as the *cost* of each type of beam, as a factor in deciding between them (since the ANOVA test shows no significant difference in their strengths).

   (c) This is similar to pat (b), except now there is no significant difference between the strengths of any of the three beams, so another factor (e.g., cost) can be used to make a decision about which one to use.

3. The two ANOVA tests will give *identical* conclusions. The reason for this is that an ANOVA test is based on comparing variances, which will not be affected by a calibration error. The calibration will certainly cause the mean of the measurements to shift, but the *variation* of the measurements around the mean will be the same as the variation of accurate measurements around the true mean. For example, if x_i denotes the *true* measured strength of an alloy sample, then $x_i + 2.5$ would be the reading given by the instrument. Letting $\bar{x}$ denote the mean of the true measurements, then $\bar{x}+2.5$ would be the mean of the instrument's measurements. Because $(x_i - \bar{x}) = (x_i + 2.5) - (\bar{x}+2.5)$, the sample *variances* of the true measurements and the instrument's measurements will be identical.

5. Using this method, there is no way of knowing whether or not there is a statistically significant difference between the means (because the within-samples variation is not used in the 'pick the winner' strategy). Consider what happens when there is actually *no* significant difference between the means (assuming an ANOVA test was conducted). This information would not be available from the pick the winner' strategy so, unlike the answer to Exercise 1 above, you would not know that you were free to use *other* factors to choose between the populations based on other criteria (e.g., cost, time, etc.).

7. We will use a to denote the upper-tail area of the F curve (e.g., $F_{.05}$ denotes the F value that is exceeded by the upper 5% of the F distribution). Then, as shown in Exercise 6, $F_{.05}(df_1 = 5, df_2 = 8) \neq F_{.05}(df_1 = 8, df_2 = 5)$ and $F_{.01}(df_1 = 5, df_2 = 8) \neq F_{.01}(df_1 = 8, df_2 = 5)$. The point is that the critical values of the F distributions depend upon the *order* as well as the numerical values of df_1 and df_2. Mistakenly reversing the order of df_1 and df_2 will give incorrect F-critical values, which may lead to incorrect ANOVA test conclusions.

9. Using $\alpha = .05$, $df_1 = 3$, and $df_2 = 20$, a critical 5% upper-tail value of F = 3.10 is found in Table VIII. Since the calculated F = 4.12, which exceed the critical value of 3.10, H_0 is rejected and we conclude that is a difference among the four means.

Section 9.2

11. (a) Using the known relationships between the entries in the ANOVA table, we can fill in the table as follows:
We know $k = 6$, $n_1 = n_2 = n_3 = n_4 = n_5 = n_6 = 26$, and $n = 156$.
The degrees of freedom are:
df associated with SSTr is $(k - 1) = (6 - 1) = 5$
df associated with SSE is $(n - k) = (156 - 6) = 150$
df associated with SST is $(n - 1) = (156 - 1) = 155$

Since MSE = 13.929 and SST = 5664.415, we know:

SSE = (MSE)(n – k) = (13.929)(150) = 2089.35
SSTr = SST – SSE = 3575.065
MSTr = SSTr/(k-1) = 3575.065/5 = 715.013, and
F = MSTr/MSE = 715.013/13.929 = 51.333

Source	Df	Sum of Squares	Mean Square	F
Mixture	5	3575.065	715.013	51.333
Error	150	2089.35	13.929	
Total	155	5664.415		

(b) H_0: There is no difference between the 6 low-permeability mixtures with regard to true average electrical resistivity; i.e., $\mu_1 = \mu_2 = \mu_3 = \mu_4 = \mu_5 = \mu_6$ where μ_i = true average electrical resistivity of the ith mixture.
H_a: At least one of the mixtures is different with regard to true average electrical resistivity; i.e., at least one of the μ_i is different.

(c) Using Minitab the P-value = $P(F_{5,150} > 51.33) \approx 0$, < 0.05. Hence, we have very strong evidence that at least one of the mixtures is different with regard to true average electrical resistivity.

13. (a) $H_0: \mu_1 = \mu_2 = \mu_3 = \mu_4 = \mu_5$ where μ_i = true average shear bond strength for i^{th} contamination/cleaning protocol.
H_a: At least one of the μ_i is different.

(b) Using the formulae from section 9.2, we can construct the ANOVA table as:
We know $k = 5$, $n_1 = n_2 = n_3 = n_4 = n_5 = 10$, and $n = 50$.
The degrees of freedom are:
df associated with SSTr is $(k - 1) = (5 - 1) = 4$
df associated with SSE is $(n - k) = (50 - 5) = 45$
df associated with SST is $(n - 1) = (50 - 1) = 49$

Overall mean = 14.52

SSTr = $10\,[(10.5 - 14.52)^2 + (14.8 - 14.52)^2 + (15.7 - 14.52)^2 + (10 - 14.52)^2 + (21.6 - 14.52)^2\,] = 881.88$
➔MSTr = 881.88/4 = 220.47

SSE = $(10\text{-}1)\,[4.5^2 + 6.8^2 + 6.5^2 + 6.7^2 + 6^2] = 1706.67$
➔MSE = 1706.67/45 = 37.93 ➔ F = MSTr/MSE = 5.813

Source	Df	Sum of Squares	Mean Square	F
Treatment	4	881.88	220.47	5.813
Error	45	1706.67	37.93	
Total	49	2588.55		

Using Minitab the P-value = $P(F_{4, 45} > 5.813)$ is 0.0007, which is < 0.01. Hence, we have very strong evidence that at least one of the contamination or cleaning protocols has a different true average shear bond strength.

15. $H_0: \mu_1 = \mu_2 = \mu_3 = \mu_4$ where μ_i = true average foam density for i^{th} manufacturer.
H_a: At least one of the μ_i is different.

Using Minitab, we find the ANOVA table to be:

One-way ANOVA: foam density versus manuf

Source	DF	SS	MS	F	P
manuf	3	15.60	5.20	2.31	0.218
Error	4	8.99	2.25		
Total	7	24.59			

Using One-Way ANOVA in Minitab, we found the F test statistic to be 2.31 and the P-value to be 0.218 > 0.05. Do not reject H_0. Hence, we do not have evidence that at least one of the manufacturers has a different true average foam density.

17. (a) Changing units of measurement amounts to simply multiplying each observation by an appropriate conversion constant, c. In this exercise, c = 2.54. Next, note that replacing each x_i by cx_i causes any sample mean to change from $\bar{\bar{x}}$ to $c\bar{\bar{x}}$ while the grand mean also changes from $\bar{\bar{x}}$ to $c\bar{\bar{x}}$. Therefore, in the formulas for SSTr and SSE, replacing each x_i by cx_i will introduce a factor of c^2. That is, SSTR(for the cx_i data) = $n_1(c\bar{x}_1 - c\bar{\bar{x}})^2 + \ldots + n_k(c\bar{x}_k - c\bar{\bar{x}})^2 = n_1c^2(\bar{x}_1 - \bar{\bar{x}})^2 + \ldots + n_kc^2(\bar{x}_k - \bar{\bar{x}})^2 = (c^2)$SSTr(for the original x_i data). The same thing happened for SSE; i.e., SSE(for the cx_i data) = (c^2)SSE(for the original x_i data). Using these facts, we also see that SST(for the cx_i data) = SSTR(for the cx_i data) + SSE(for the cx_i data) = (c^2)SSTr(for the original x_i data)+ (c^2)SSE(for the original x_i data) = (c^2)[SSTR(for the original x_i data)+SSE(for the original x_i data)] = (c^2)SST(for the original x_i data). The net effect of the conversion, then, is to multiply all the sums of square in the ANOVA table by a factor of $c^2 = (2.54)^2$. Because neither the number of treatments nor the number of observations is altered, the entries in the degrees of freedom column of the ANOVA table is not changed. Notice also that the F-ratio remains unchanged: F(for the cx_i data) = MSTR(for the cx_i data)/MSE(for the cx_i data) = (c^2)MSTR(for the original x_i data)/[$(c)^2$MSE(for the original x_i data)] = MSTR(for the original x_i data)/MSE(for the original x_i data) = F-ratio(for the original x_i data). This makes sense, for otherwise we could change the significance of an ANOVA test by merely changing the units of measurement.

(b) The argument in (a) holds for *any* conversion factor c, not just for c = 2.54. We can conclude then, that *any* change in the units of measurement will change the 'Sum of Squares' column in the ANOVA table, but the degrees of freedom and F ratio will remained unchanged.

19. Since the sample sizes are equal, we can simply average the three sample means to find the grand mean $\bar{\bar{x}}$ = (3.43+3.18+3.22)/3 = 3.2767. Next, SSE = $(n_1-1)s_1^2 + (n_2-1)s_2^2 + (n_3-1)s_3^2 = (6-1)(.22)^2 + (6-1)(.13)^2 + (6-1)(.11)^2 = .3870$ and SSTr = $n_1(\bar{x}_1 - \bar{\bar{x}})^2 + n_2(\bar{x}_2 - \bar{\bar{x}})^2 + n_3(\bar{x}_3 - \bar{\bar{x}})^2 = 6(3.43-3.2767)^2 + 10(3.18-3.2767)^2 + 10(3.22-3.2767)^2 = .21640$. Since k = 3 and n = (3)(6) = 18, SST has d.f. = n - 1 = 18-1 = 17, SSTR has d.f. = k-1 = 3- 1 = 2, and SSE has d.f. - n-k = 18-3 = 15. Therefore, MSTr = SSTr/(k-1) = .21640/2 = .10820 and MSE = SSE/(n-k) = .3870/15 = .0258. The test statistic is F = MSTr/MSE = .10820/.0258 = 4.194. Since the value of F = 4.194 falls between the values 3.68 and 6.36 in Table VIII (for $df_1 = 2$, $df_2 = 15$), we can conclude that .01 < P-value < .05. Since the P-value is smaller than $\alpha = .05$, H_0 is rejected and we conclude that there is a difference between the means for the three age categories.

21. (a) The relevant hypotheses are H_0: $\mu_0 = \mu_{45} = \mu_{90}$ versus H_a: *at least two of the population means differ*. The ANOVA table from a Minitab printout appears below (*note: we entered the data into columns c1, c1, and c3, and used the 'unstack' ANOVA command*).

Analysis of Variance

Source	DF	SS	MS	F	P
Factor	2	25.80	12.90	2.35	0.120
Error	21	115.48	5.50		
Total	23	141.28			

Minitab computes the exact P-value, but we can still approximate it by using $df_1 = 2$ and $df_2 = 21$ in Table VIII. In Table VIII, F = 2.35 is smaller than any of the entries in the table (for $df_1 = 2$, $df_2 = 21$), so we know that the P-value > .10. Since P-value > .10 > .05 = α, H_0 is not rejected and we conclude that there is no evidence of a difference between the mean strengths at the three different orientations.

(b) The results in (a) are favorable for the practice of using wooden pegs because the pegs do not appear to be sensitive to the angle of application of the force. Since the angle at which force is applied could vary widely at construction sites, it is good to know that peg strength is not seriously affected by different construction practices (i.e., possibly different force application angles).

23. (a) Let x_i denote the true value of an observation. Then x_i +c is the measured value reported by an instrument which is consistently off (i.e., out of calibration) by c units. Therefore, if $\bar{x}$ denotes the mean of the true measurements, then $\bar{x}$+c is the mean of the measured values. Similarly, the grand mean of the measured values equals $\bar{\bar{x}}$ + c, where $\bar{\bar{x}}$ is the grand mean of the true values. Putting these results in the formula for SSTR, we find SSTr(measured values) = $n_1(\bar{x}_1 + c - (\bar{\bar{x}} + c))^2 + \ldots + n_k(\bar{x}_k + c - (\bar{\bar{x}} + c))^2 = n_1(\bar{x}_1 - \bar{\bar{x}})^2 + \ldots + n_k(\bar{x}_k - \bar{\bar{x}})^2$ = SSTR (true values). Furthermore, note that any sample variance is unchanged by the calibration problem since the deviations from the mean for the measured data are identical to the deviations from the mean for the true values; i.e., $(x_i +2.5) - (\bar{x} +2.5) = (x_i - \bar{x})$. Therefore, SSE (measured values) = $(n_1-1)s_1^2 + \ldots + (n_k-1)s_k^2$ = SSE (true values). Finally, because SSTR and SSE are unaffected, so too will SST be unaffected by the calibration error since SST = SSTR + SSE. Thus, *none* of the sums of squares are changed by the calibration error. Obviously, the degrees of freedom are unchanged too, so the net result is that there will be no change in the entire ANOVA table.

(b) Calibration error will not change any of the ANOVA table entries and therefore will not affect the results of an ANOVA test. That is, if all data points are shifted (up or down) by the same amount c, the ANOVA entries will not be affected. However, the mean of each sample *will* shift by an amount equal to c.

Section 9.3

25. An effects plot only shows the sample means. It does not show the within-sample variance, so the variation between groups can not be compared to the within-group variation.

27. The sample sizes for the k = 5 brands are equal, ($n_1 = n_2 = n_3 = n_4 = n_5 = 4$), so the appropriate degrees of freedom for the Studentized range are k = 5 and n-k = $n_1+n_2+n_3+n_4+n_5$ = 20 - 5 = 15. Using a significance level of α = .05, we find $q_{.05}(5,15) = 4.37$ from Table IX, so the threshold value is $T = q_{.05}\sqrt{\frac{MSE}{n_i}} = (4.37)\sqrt{\frac{272.8}{4}} = 36.09$. Arranging the sample means from smallest to largest, the means that are no further than T = 36.09 apart are underlined:

437.5 462.0 469.3 512.8 532.1
(underlined: 437.5 – 469.3; 512.8 – 532.1)

Two of the brands appear to have significantly higher average coverage areas than the other three.

29. The value of T = 36.09 will be the same as in Exercise 27. After arranging the sample means from smallest to largest, the ones that are closer than T = 36.09 are underlined:

427.5 462.0 469.3 502.8 532.1
(underlined: 427.5 – 462.0; 462.0 – 469.3; 469.3 – 502.8; 502.8 – 532.1)

The conclusion is similar to that in Exercise 27, but now only one of the brands (the one with mean 532.1) has significantly higher coverage areas than the three with the smallest means.

31. We know that k = 6 and $n_i = 26$, so the critical value is $q_{.05,6,150} \approx q_{.05,6,120} = 4.10$, and MSE = 13.929. So, $T = 4.10\sqrt{\frac{13.929}{26}} = 3.00$. So, sample means less than 3.00 apart are not statistically significantly different, and will be joined by a line. Three distinct groups emerge: the first mixture, then mixtures 2-4, and finally mixtures 5-6.

14.18 17.94 18.00 18.00 25.74 27.67
(underlined: 17.94 – 18.00; 25.74 – 27.67)

33. We can use Minitab (Stat > One-Way ANOVA > Comparisons > Tukey) to obtain the following results, at $\alpha = 0.05$:

```
Grouping Information Using Tukey Method
pulse  N    Mean  Grouping
120    3  53.667  A
100    3  43.333    B
140    3  42.667    B

Means that do not share a letter are significantly different.
```

Which can be re-expressed as:

42.667 43.333 53.667
(underlined: 42.667 – 43.333)

Thus, at a 5% level of significance, we can conclude that a pulse of 120 has a higher average toughness compared to a pulse of 100 or 140.

Section 9.4

35. (a) The following Minitab ANOVA table was created using the command *MTB> anova c1 = c2 c3* (where the data is in column c1; row and column subscripts are in c2 and c3, respectively):

Source	DF	SS	MS	F	P
brand	4	53231	13308	95.57	0.000
level	3	116218	38739	278.20	0.000
Error	12	1671	139		
Total	19	171120			

The F-ratio for '*brand*' is F = 95.57. For $df_1 = 4$ and $df_2 = 12$, the value F = 95.57 has a P-value smaller than .001 (From Table VIII). Since this P-value is smaller than $\alpha = .01$, we can conclude that there is a difference in power consumption among the 5 brands

(b) The F-ratio for '*humidity*' is F = 278.20. For $df_1 = 3$ and $df_2 = 12$, the P-value associated with F = 278.20 is less than .001 (from Table VIII), so the null hypothesis H_0: *average humidity is the same* is rejected. We conclude that humidity levels do affect power consumption, so it was wise to use humidity as a blocking factor.

37. The following ANOVA output was obtained from Minitab using the command *MTB> anova c1 = c2 c3* (where the data is in column c1; row and column subscripts are in c2 and c3, respectively):

```
Analysis of Variance for %accept, using Adjusted SS for Tests
```

Source	DF	Seq SS	Adj SS	Adj MS	F	P
operator	2	27.556	27.556	13.778	10.78	0.024
brand	2	22.889	22.889	11.444	8.96	0.033
Error	4	5.111	5.111	1.278		
Total	8	55.556				

(a) The F- ratio for '*brand*' is F = 8.96. For $df_1 = 2$ and $df_2 = 4$, the P-value associated with F = 8.96 is 0.033 (Table VIII). Since this P-value is less than $\alpha = .05$, we reject H_0:*no difference in average % acceptable among lathe brands* and conclude that there is evidence that the type of lathe brand does have an effect on % acceptable.

(b) The F- ratio for '*operator*' is F = 10.78. For $df_1 = 2$ and $df_2 = 4$, the P-value associated with F = 10.78 is 0.024(Table VIII). Since this P-value is less than $\alpha = .05$, we reject H_0:*no difference in average % acceptable among operators* and conclude that there is evidence that the different operators have differing effects on product acceptability. So, *blocking* was a good idea.

39. (a) The ANOVA table can be completed as follows:
We n = 4*21 = 84
The degrees of freedom are:
df associated with Design = 4-1 = 3
df associated with Person = 21-1 = 20
df associated with SST is (n – 1) = (84 – 1) = 83
df associated with Residuals = 83 – 3 - 20 =60

MSDesign = SSDesign/3 = 519515/3 = 173171.667
MSE = SSE/60 = 293009/60 = 4883.48333
SSPerson = 20*MSPerson = 20*5023 = 100460

F_{design} = MSDesign/MSE = 173171.667/4883.4833 = 35.4606855
F_{person} = MSPerson/MSE = 5023/4883.48333 = 1.02856909
The P-value corresponding to Design can be found using Minitab, and is $P(F_{3, 60} > 35.4607) \approx 0 < 0.0001$.

Source	Df	Sum of Squares	Mean Square	F	P-value
Design	3	519515	173171.667	35.4606855	<0.0001
Person	20	100460	5023	1.02856909	0.445
Error	60	293009	4883.48333		
Total	83				

(b) The F- ratio for '*design*' is F = 35.46. For $df_1 = 3$ and $df_2 = 60$, the P-value associated with F = 35.46 is ≈ 0 (Table VIII). Since this P-value is less than $\alpha = .05$, we reject H_0:*no difference in average RPN among designs* and conclude that there is evidence that the type of design does have an effect on RPN.

(c) The F- ratio for '*person*' is F = 1.03. For $df_1 = 20$ and $df_2 = 60$, the P-value associated with F = 1.03 is 0.445 (given). Since this P-value is not less than $\alpha = .05$, we do not reject H_0:*no difference in average RPN among persons* and conclude that there is no evidence that the RPN differs from person to person.

41. (a) The following Minitab ANOVA table was created using the command *MTB> anova c1 = c2 c3* (where the data is in column c1; row and column subscripts are in c2 and c3, respectively):

```
Analysis of Variance for strength

Source      DF        SS        MS       F      P
batch        9    86.793     9.644    7.22  0.000
method       2    23.229    11.614    8.69  0.002
Error       18    24.045     1.336
Total       29   134.067
```

The F- ratio for 'method' is F = 8.69. For $df_1 = 2$ and $df_2 = 18$, the P-value associated with F = 8.69 is between .001 and .01 (Table VIII). Since this P-value is less than $\alpha = .05$, we reject H_0: *no difference in average strength among the different methods* and conclude that curing methods do have differing effects on strength.

(b) The F-ratio for 'batch' is F = 7.22. For $df_1 = 9$ and $df_2 = 18$, the P-value associated with F = 7.22 is less than .001. Since this P-value is less than $\alpha = .05$, we reject H_0: *no difference in average strength between different batches* and conclude that different batches do have an effect on strength.

(c) Ignoring '*batch'* and simply conducting a single factor ANOVA using '*method'*, we obtain the following Minitab output (we used the command MTB> *anova c1 =c2*):

```
Analysis of Variance for strength

Source      DF        SS        MS       F      P
method       2    23.229    11.614    2.83  0.077
Error       27   110.838     4.105
Total       29   134.067
```

The F-ratio for '*method*' is now F = 2.83. For $df_1 = 2$ and $df_2 = 27$, the P-value associated with F = 2.83 is between 5% and 10%, i.e., .05 < P-value < .10. Since this P-value is not smaller than $\alpha = .05$, we would not reject H_0: *no difference in average strength among curing methods* and conclude that curing method does not affect strength. This result is caused by the fact that in this single-factor ANOVA we have ignored differences between batches (whose effects have been incorporated into the MSE which leads to a smaller F-ratio than in part (a)). That is, ignoring an important blocking factor (*'batch'*) can obscure the differences between levels of the factor you are interested in (*'curing method'*).

Supplementary Exercises

43. (a) Let μ_i denote the true mean lumen output for light bulb brand i where $i = 1, 2, 3$. The relevant hypotheses are:

H_0: $\mu_1 = \mu_2 = \mu_3$ versus
H_a: at least two of the population means differ.

(b)

Source	df	SS	MS	F
Factor	2	591.2	295.6	1.3
Error	21	4773.3	227.3	
Total	23	5364.5		

(c) The F-ratio is 1.30. The corresponding P-value = P(F > 1.30). Using Table VIII with $df_1 = 2$ and $df_2 = 21$, we obtain an estimated P-value > .10. Since the P-value > α = .05, we do not reject H_0. We do not have enough evidence to conclude that there are differences between the average lumen outputs for the three brands.

45. (a) From Table VIII, $F_{.05}(1,10) = 4.96$ and $t_{.025}(10) = 2.228$ and $(2.228)^2 \approx 4.96$. The equality is approximate because the entries in the F and t table entries are rounded.

(b) $F_\alpha(df_1=1, df_2) = (t_{\alpha/2})^2$, so for $\alpha = .05$: $F_{.05}(1, df_2) = (t_{05/2})^2 = (t_{.025})^2$, which approaches $(z_{.025})^2 = (1.96)^2 = 3.8416$.

47. $n_1 = n_2 = n_3 = 5$, so k = 3 and $n = n_1+n_2+n_3 = 15$. Because the sample sizes are equal, the grand mean is simply the average of the three sample means, $\bar{\bar{x}} = (10+12+20)/3 = 14$. Therefore, $SSTr = n_1(\bar{x}_1 - \bar{\bar{x}})^2 + n_2(\bar{x}_2 - \bar{\bar{x}})^2 + n_3(\bar{x}_3 - \bar{\bar{x}})^2 = 5(10-14)^2 + 5(12-14)^2 + 5(20-14)^2 = 280$. Then, MSTR = SSTR/(k-1) = 280/(3-1) = 140 and MSE = SSE/(n-k) = SSE/(15-3) = SSE/12, so the F-ratio is F = MSTR/MSE = 140/MSE = 140/(SSE/12). = 1680/SSE. For $df_1 = 2$ and $df_2 = 12$, the value of F associated with a right-tail area $\alpha = .05$ is F = 3.89. So, to reject H_0 (Condition 1), we must have F = 1680/SSE > 3.89 or, 1680/3.89 = 431.88 > SSE.

For Condition 2, we first look up $q_{.05}(k, n\text{-}k) = q_{.05}(3,12) = 3.77$ in Table IX. Next, $T = q_{.05}\sqrt{\frac{MSE}{n_i}} = (3.77)\sqrt{\frac{SSE/12}{5}}$. Therefore, if none of the three sample means are to be further than T units apart, the *largest* of differences between sample means (i..e, 20 - 10 = 10) must be smaller than T. That is $10 < T = (3.77)\sqrt{\frac{SSE/12}{5}}$. Solving for SSE, we find SSE > $(10/3.77)^2(60) = 422.16$.

Therefore, for <u>both</u> Conditions 1 and 2 to hold, we must have 422.16 < SSE < 431.88.

49. Use the same thinking as in Exercise 47.

The grand mean equals (10 + 15 + 20)/3 = 15. Therefore, SSTr = 250, MSTr = 125, and MSE = $\left(\frac{SSE}{12}\right)$.

So, F = $\left(\frac{MSTr}{MSE}\right) = (125)\left(\frac{12}{SSE}\right) = \left(\frac{1500}{SSE}\right)$.

For $df_1 = 2$ and $df_2 = 12$ and $\alpha = .05$, the F critical value = 3.89.

So, to reject H_0 (condition 1), we must have F = $\left(\frac{1500}{SSE}\right) > 3.89$.

Or, 385.6 > SSE.

For condition 2, we first look up $q_{.05}(3, 12) = 3.77$. Next,

$$T = (q_{.05})\left(\sqrt{\frac{MSE}{n_i}}\right) = (3.77)\left(\sqrt{\frac{\left(\frac{SSE}{12}\right)}{5}}\right)$$

So, the largest of the differences between sample means (i.e., 20 – 10 = 10) must be smaller than T for condition 2 to be met.

Solving for SSE, we find:

SSE > $\left(10/3.77\right)^2(60) = 422.15$.

Therefore, both condition 1 (SSE < 385.60) and condition 2 (SSE > 422.15) cannot be met.

51. (a)

Source	df	SS	MS	F	P-value
Drying method	4	14.962	3.741	36.70	0.000
Fabric type	8	9.696	1.212	11.89	0.000
Error	32	3.262	0.102		
Total	44	27.920			

(b) The null hypothesis of interest is H_0: there are no differences in mean smoothness scores for the five drying methods.

The F-ratio for "drying method" is F = 36.7. For $df_1 = 4$ and $df_2 = 32$, the P-value reported by Minitab is 0.000. The P-value estimated using Table VIII is < .001. Since the P-value is less than $\alpha = .05$, we reject H_0. We have sufficient evidence to conclude that there are differences in mean smoothness scores for the five drying methods.

Chapter 10

Experimental Design

Section 10.1

1. Replication allows you to obtain an estimate of the *experimental error*, which is sometimes thought of as the "noise". That is, we expect a certain amount of natural variation between experimental results, even when all factors are held fixed, and we call this variation the experimental error. Knowing the magnitude of the experimental error allows you to know when a factor's effect is important or not; important/significant factors are those whose effect on the response variable causes changes/variation that is larger in magnitude than the experimental error.

3. (a) and (b): The following surface plot of the function f(x,y) was created in MathCAD. The surface is a dome whose maximum point sits over the point x = 2, y = 5 in the x-y plane:

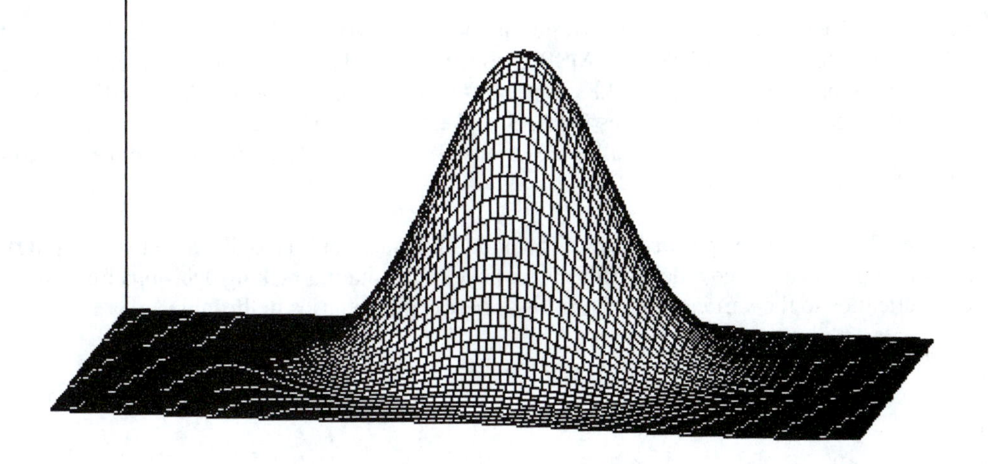

(c) For a contour level of c, we set f(x,y) = c. Taking natural logarithms of both sides of the equation $e^{-\frac{1}{2}\left[(x-2)^2+(y-5)^2\right]} = c$, we find $-\frac{1}{2}\,[(x-2)^2+(y-5)^2] = \ln(c)$. Note that c must be positive (because the exponential function is always positive) and less than 1 (since the expression in the exponent is always negative and the exponential function is less than 1 for negative arguments). Thus, ln(c) must be negative. Multiplying through by -2 then yields the familiar equation $(x-2)^2+(y-5)^2 = -2\ln(c)$, where -2ln(c) is a positive number. This is the equation of a circle in the plane with radius equal to the square root of -2ln(c). That is, all contours for f(x,y) are circles in the plane.

(d) Sketching the contours should show that the maximum is achieved when the expression in the exponent of f(x,y) is 0. Because $(x-2)^2+(y-5)^2$ is nonnegative, it can only equal 0 when both x = 2 and y = 5. Another method of obtaining the maximum point would be to take partial derivatives of f(x,y), set them equal to 0, and solve. For example, the equations

$$\frac{\partial}{\partial x}\, e^{-\frac{1}{2}\left[(x-2)^2+(y-5)^2\right]} = -(x-2)\, e^{-\frac{1}{2}\left[(x-2)^2+(y-5)^2\right]} = 0 \quad \text{and}$$

$$\frac{\partial}{\partial y}\, e^{-\frac{1}{2}\left[(x-2)^2+(y-5)^2\right]} = -(y-5)\, e^{-\frac{1}{2}\left[(x-2)^2+(y-5)^2\right]} = 0$$

have the unique solution x = 2, y =5.

Section 10.2

5. When the lines in the AB interaction plot are parallel, the effect of changing Factor A (or Factor B) from one level to another will be the same for each fixed level of Factor B (or Factor A). That is, when factors A and B do not interact, their effect on the response variable (i.e., the amount that they change the response variable) does not depend on the particular level of the other factor. Since the slope of a line between two values of Factor A is proportional to the *change* in the response, and since this change is not affected by the levels of B, the slope will be the same (i.e., the lines will be parallel) for any fixed level of factor B.

7. Each factor has 5 levels, so each has d.f. = 5-1 = 4. The total number of experimental runs is n = abr = (5)(5)(3), so the total d.f. = abr-1 = 75 - 1 = 74. The AB interaction term has df = (a-1)(b-1) = (5-1)(5-1) = 16. The degrees of freedom for error can then be found by subtracting the factor and interaction degrees of freedom from the total; i.e.; error d.f. = 74 - 4 - 4 - 16 = 50. Alternatively, you could use the formula: error d.f. = ab(r-1) = (5)(5)(3-1) = 50.

Next, MSA = SSA/(a-1) = 20/4 = 5, so the F-ratio for Factor A = MSA/MSE = 5/2 = 2.5. Similarly, the F-ratio for Factor B is 8.1 = F = MSB/MSE = MSB/2, so MSB = 2(8.1) = 16.2 and therefore, 16.2 = MSB = SSB/(b-1) = SSB/4 or, SSB = 4(16.2) = 64.8. Since 2 = MSE = SSE/50, we also have SSE = 2(50) = 100. To find SS(AB), just subtract the SS values for A, B, and error from SST; i.e.; SS(AB) = SST - SSA - SSB - SSE = 200 - 20 - 64.8 - 100 = 15.2. Finally, MS(AB) = SS(AB)/16 = 15.2/16 = .95 and the F-ratio for AB is F = MS(AB)/MSE = .95/2 = .475.

The completed ANOVA table appears below. The point of this exercise is to illustrate that an ANOVA table contains a great deal of redundancy (for the purpose of facilitating making decisions from the table) and to draw attention to the various interrelationships between the entries in the table.

Source	df	SS	MS	F
Factor A	4	20	5	2.5
Factor B	4	64.8	16.2	8.1
Interaction	16	15.2	0.95	0.475
Error	50	100	2	
Total	74	200		

9 (a) Let Factor A be 'formulation' and let Factor B be 'speed'. Then a = 2, b = 3, and the number of replications shown in the data array is r = 3. Therefore, the error d.f. = ab(r-1) = (2)(3)(3-1) = 12, so MSE = SSE/12 = 71.87/12 = 5.9892. MS(AB) = SS(AB)/[(a-1)(b-1)] = 18.58/[(2-1)(3-1)] = 9.29, so the F-ratio for the interaction term is then F = MS(AB)/MSE = 9.29/5.9892 = 1.551. Using df_1 = (a-1)(b-1) = 2 and df_2 = ab(r-1) = 12, the P-value associated with F = 1.551 is greater than .10. Therefore, H_0:*there is no AB interaction effect* can not be rejected. That is, we have no evidence of any interaction between factors A and B.

(b) The F-ratio for Factor A is F = MSA/MSE = [SSA/(a-1)]/MSE = [2253.44/(2-1)]/ 5.9892 = 376.25. Using df_1 = (a-1) = 1 and df_2 = ab(r-1) = 12, the P-value associated with F = 376.25 is less than .001 (Table VIII), so at significance level α = .05 we can reject H_0:*there is no effect for Factor A* and conclude that different chemical formulations do have an effect on yield.

The F-ratio for Factor B is F = MSB/MSE = [SSB/(b-1)]/MSE = [230.81/(3-1)]/ 5.9892 = 19.269. Using df_1 = (b-1) = 2 and df_2 = ab(r-1) = 12, the P-value associated with F = 19.269 is less than .001 (Table VIII), so at significance level α = .05 we can reject H_0:*there is no effect for Factor B* and conclude that different speeds also have an effect on yield.

11. The following Minitab ANOVA output was obtained:

```
Analysis of Variance for cond, using Adjusted SS for Tests

Source                       DF     Seq SS     Adj SS     Adj MS      F      P
Binder grade                  2  0.0020893  0.0020893  0.0010447  14.12  0.002
Agg content                   2  0.0082973  0.0082973  0.0041487  56.06  0.000
Binder grade*Agg content      4  0.0003253  0.0003253  0.0000813   1.10  0.414
Error                         9  0.0006660  0.0006660  0.0000740
Total                        17  0.0113780
```

(a) H_o: There is no interaction effect of binder grade and aggregate content
Since the P-value corresponding to the interaction is 0.414 which is > $\alpha = .01$, we do not reject H_0. We conclude that there is no evidence of an interaction between the two factors.

(b) H_o: There is no main effect for factor aggregate content.
Since the P-value corresponding to Aggregate content is less than 0.001 (which is less than $\alpha = .01$), we reject H_0. We conclude that aggregate content does affect thermal conductivity.

(c) H_o: There is no main effect for factor Binder grade.
Since the P-value corresponding to factor Binder grade is 0.002 (which is less than $\alpha = .01$), we reject H_0. We conclude that aggregate content does have an effect on thermal conductivity.

13. (a) Given below is a set of boxplots generated using Minitab. SBP-dry appears to have the highest mean (median) shear bond strength, with OBP-moist showing the highest variability. Only OBP-dry appears to have an outlier.

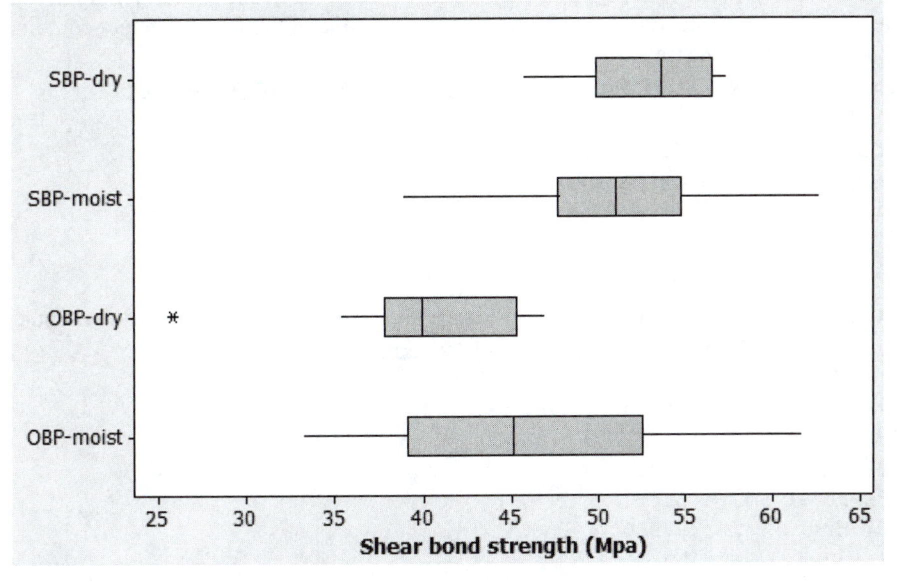

(b) Using Minitab (see output below), we obtained the following ANOVA output. We find that only type of adhesive has a P-value < 0.001 which is less than 0.01. Hence, we have strong evidence that there is a difference in true mean shear bond strengths of the two adhesive types. However, we do not have any evidence of an interaction (P-value = 0.031 which is not less than 0.01), or evidence of an effect of condition (P-value = 0.288 which is not less than 0.01).

```
Analysis of Variance for SBS(MPa), using Adjusted SS for Tests

Source           DF   Seq SS   Adj SS   Adj MS      F      P
adhesive          1   951.41   951.41   951.41  22.85  0.000
condn             1    48.20    48.20    48.20   1.16  0.288
adhesive*condn    1   207.92   207.92   207.92   4.99  0.031
Error            44  1832.32  1832.32    41.64
Total            47  3039.85
```

The following are main effects and interaction plots.

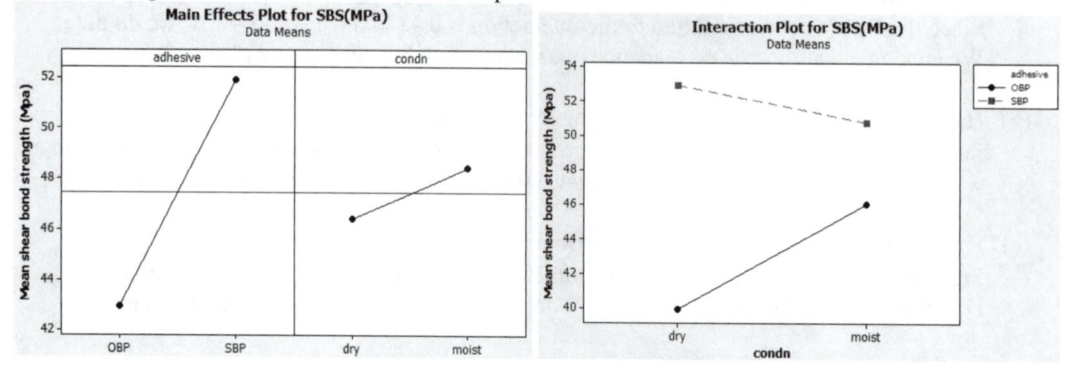

(c) Using One-Way ANOVA in Minitab (see output below), we find that the P-value is less than 0.001 which is less than 0.01. We reject H_0: there is no difference in mean shear strength of the four treatment combinations. Hence, we have strong evidence that at least one of the four treatment combinations has a different mean shear bond strength.

One-way ANOVA: SBP-dry, SBP-moist, OBP-dry, OBP-moist

```
Source  DF      SS     MS     F      P
Factor   3  1207.5  402.5  9.67  0.000
Error   44  1832.3   41.6
Total   47  3039.8
```

Using Tukey's method at an overall 5% significance level we obtained the following output:

```
Grouping Information Using Tukey Method

             N    Mean  Grouping
SBP-dry     12  52.967  A
SBP-moist   12  50.808  A
OBP-moist   12  46.067  A B
OBP-dry     12  39.900    B

Means that do not share a letter are significantly different.
```

This can be also presented as follows, arranging the sample means from smallest to largest, and joining those means by a line between which no statistically significant difference was found:

39.9 46.067 50.08 52.967

We have evidence at an overall 5% significant level that SBP-Dry and SBP-Moist each have higher average shear bond strengths than OBP-Dry. We found no evidence of a difference between the means of the other pairs of treatment combinations.

Section 10.3

15. (a) We know that MS = SS/df; $F_{calculated} = MS_{factor}/MSE$; and the corresponding P-value = $P(F_{df1, df2} > F_{calculated})$

Also we know that:
df for A = 3-1 = 2; df for B = 3-1 = 2; df for C = 3-1 = 2;
df for AB = (3-1)x(3-1) = 4;
df for AC = (3-1)x(3-1) = 4;
df for BC = (3-1)x(3-1) = 4;
df for ABC = (3-1)x(3-1)x(3-1) = 8;
df for Total = 3*3*3*2 – 1 = 53;
df for Error = 53 – 26 = 27

Source	df	SS	MS	F	P-value (from Minitab)
A	2	210.67	105.335	0.525	0.5975
B	2	132.17	66.085	0.329	0.7225
C	2	2586.35	1293.175	6.446	0.0051
AB	4	57.48	14.37	0.072	0.9900
AC	4	636.84	159.21	0.794	0.5394
BC	4	875	218.75	1.090	0.3813
ABC	8	888.52	111.065	0.554	0.8053
Error	27	5416.67	200.6174		
Total	53	10803.70			

(b) None of the interaction effects are significant; the corresponding P-values for each of the interactions (ABC, AB, AC, BC) are > 0.05. (Also, the corresponding F critical value is 2.728 for two-factor interactions, and 2.305 for three-factor interactions, which are larger than the corresponding calculated F statistic values for the interactions.)

(c) The main effect of quill gap (C) is significant, because the P-value = 0.0051 < 0.05. (Also, the corresponding F critical value is 3.354.) Thus, we have evidence that at least one of the quill gaps has a different true average surface roughness.

17. (a) Each factor has three levels: a = 3, b = 3, c = 3. Therefore, the degrees of freedom for each factor is a-1 = b-1 = c-1 = 2; the degrees of freedom for each 2-factor interaction is (a-1)(b-1) =(a-1)(c-1) = (b-1)(c-1) = 4; and the 3-factor interaction has d.f. = (a-1)(b-1)(c-1) = 8. Using these d.f., the sums of squares are converted to mean squares and F-ratios in the ANOVA table:

Source	df	SS	MS	F
A	2	14,144.44	7,072.22	61.06
B	2	5,511.27	2,755.64	23.79
C	2	244,696.39	2,348.20	1,056.27
AB	4	1,069.62	267.20	2.31
AC	4	62.67	15.67	0.14
BC	4	311.67	82.92	0.72
ABC	8	1,080.77	135.10	1.17
Error	27	3,127.50	115.83	
Total	53	270,024.33		

(b) The 2-factor interactions are based on $df_1 = 4$ and $df_2 = 27$. From Table VIII, the P-values of the AB, AC, and BC F-ratios all exceed .10. At a significance level of $\alpha = .05$. none of these P-values are significant; i.e., none of the 2-factor interactions are important.

(c) F-ratios for the main effects (A, B, and C) are based on $df_1 = 2$ and $df_2 = 27$. From Table VIII, the P-values of all three F-ratios are less than .001. Using a significance level of $\alpha = .05$, all three main effects are significant.

19. (a) The numbers of factors levels are: a= 3, b = 2, and c = 4. A Minitab printout of the ANOVA table for this experiment is shown below (*note: In Minitab, put all the data in one column, say c10, and then put the "subcripts" (i.e., factor levels) in columns c1, c2, and c3 and then use the command MTB> anova c10 = c1/c2/c3; the slash marks between columns lets Minitab know to compute the various interaction terms between those columns*)

Factor	Type	Levels	Values
A	fixed	3	1 2 3
B	fixed	2	1 2
C	fixed	4	1 2 3 4

Analysis of Variance for length

Source	DF	SS	MS	F	P
A	2	12.896	6.448	1.04	0.360
B	1	100.042	100.042	16.10	0.000
C	3	393.417	131.139	21.10	0.000
A*B	2	1.646	0.823	0.13	0.876
A*C	6	71.021	11.837	1.90	0.092
B*C	3	1.542	0.514	0.08	0.969
A*B*C	6	9.771	1.628	0.26	0.953
Error	72	447.500	6.215		
Total	95	1037.833			

(b) Comparing the P-values (in the ANOVA table) for the 3 main effects (A, B, and C) to $\alpha = .05$, we see that only the main effects for factors B and C have P-values less than .05, so only these effects (i.e., B and C) are important.

(c) None of the F-ratios for the interaction terms are below $\alpha = .05$, so none of the interaction terms are significant.

21. (a) Using Minitab, we obtained the following ANOVA table:

Analysis of Variance for WtNi, using Adjusted SS for Tests

Source	DF	Seq SS	Adj SS	Adj MS	F	P
Power	2	123.53	123.53	61.76	4.77	0.043
Speed	2	20.91	20.91	10.46	0.81	0.479
Thickness	2	360.55	360.55	180.28	13.92	0.002
Power*Speed	4	57.67	57.67	14.42	1.11	0.414
Power*Thickness	4	61.83	61.83	15.46	1.19	0.384
Speed*Thickness	4	11.28	11.28	2.82	0.22	0.921
Error	8	103.61	103.61	12.95		
Total	26	739.38				

(b) The P-values for the corresponding F ratios were calculated by Minitab, and from the output you can see that the P-values corresponding to the interactions (Power*Speed, Power*Thickness, and

Speed*Thickness) are all larger than 0.05. Thus, we do not have any evidence of significant two-factor interactions in the above ANOVA.

(c) At a 5% significance level, the main effects of power and thickness are significant, but not Speed, because the P-values corresponding to power (0.043) and thickness (0.002) are smaller than 0.05.

23. (a) Using Minitab, we obtained the following ANOVA table:

Analysis of Variance for permeability, using Adjusted SS for Tests

Source	DF	Seq SS	Adj SS	Adj MS	F	P
Denier	2	105793	105793	52897	1342.30	0.000
Temp	2	34436	34436	17218	436.92	0.000
Pressure	2	516398	516398	258199	6552.04	0.000
Denier*Temp	4	6868	6868	1717	43.57	0.000
Denier*Pressure	4	10178	10178	2545	64.57	0.000
Temp*Pressure	4	10922	10922	2731	69.29	0.000
Denier*Temp*Pressure	8	6713	6713	839	21.30	0.000
Error	27	1064	1064	39		
Total	53	692372				

(b) The P-values for the corresponding F ratios were calculated by Minitab, and from the output you can see that the P-values corresponding to the interactions are all much smaller than 0.01. Thus, we have strong evidence of significant two-factor and three-factor interactions in the above ANOVA.

(c) At a 5% significance level, all three main effects (Denier, Temp, and Pressure) are significant, because the P-values are all smaller than 0.01.

Section 10.4

25.

A	B	C	D	AB	AC	AD	BC	BD	CD
-1	-1	-1	-1	1	1	1	1	1	1
1	-1	-1	-1	-1	-1	-1	1	1	1
-1	1	-1	-1	-1	1	1	-1	-1	1
1	1	-1	-1	1	-1	-1	-1	-1	1
-1	-1	1	-1	1	-1	1	-1	1	-1
1	-1	1	-1	-1	1	-1	-1	1	-1
-1	1	1	-1	-1	-1	1	1	-1	-1
1	1	1	-1	1	1	-1	1	-1	-1
-1	-1	-1	1	1	1	-1	1	-1	-1
1	-1	-1	1	-1	-1	1	1	-1	-1
-1	1	-1	1	-1	1	-1	-1	1	-1
1	1	-1	1	1	-1	1	-1	1	-1
-1	-1	1	1	1	-1	-1	-1	-1	1
1	-1	1	1	-1	1	1	-1	-1	1
-1	1	1	1	-1	-1	-1	1	1	1
1	1	1	1	1	1	1	1	1	1

ABC	ABD	ACD	BCD	ABCD
-1	-1	-1	-1	1
1	1	1	-1	-1
1	1	-1	1	-1
-1	-1	1	1	1
1	-1	1	1	-1
-1	1	-1	1	1
-1	1	1	-1	1
1	-1	-1	-1	-1
-1	1	1	1	-1
1	-1	-1	1	1
1	-1	1	-1	1
-1	1	-1	-1	-1
1	1	-1	-1	1
-1	-1	1	-1	-1
-1	-1	-1	1	-1
1	1	1	1	1

Using the same process described in example 10.6, the remaining contrasts and effects are as follows:

Effect Name	Contrast	Effect
A	-33.84	-4.23
B	-6.94	-0.8675
C	3.98	0.4975
D	150.94	18.8675
AB	-13.8	-1.725
AC	-4.76	-0.595
AD	-2.56	-0.320
BC	24.62	3.0775
BD	15.42	1.9275
CD	3.26	0.4075
ABC	-1.20	-0.150
ABD	-1.408	-0.185
ACD	1.20	0.150
BCD	5.98	0.7475
ABCD	2.04	0.255

27. (a) The following ANOVA table was produced in Minitab:
Analysis of Variance for welding current

Source	df	SS	MS	F	P-value
A	1	1685.1	1685.1	102.38	0.000
B	1	21272.2	21272.2	1292.36	0.000
C	1	5076.6	5076.6	308.42	0.000
AB	1	36.6	36.6	2.22	0.174
AC	1	0.4	0.4	0.03	0.877
BC	1	109.2	109.2	6.63	0.033
ABC	1	23.5	23.5	1.43	0.266
Error	8	131.7	16.5		
Total	15	28335.3			

(b) At $\alpha = .01$, all three main effects are important, since each of their P-values is less than .001. No significant interaction effects exist, when testing at $\alpha = .01$.

29. Given the following table of data, where "-" represents the factor set at the low level, and "+" set at the high level:

Run	Storage time	Storage temp	Packaging	Color Quality Replication 1	Color Quality Replication 2
1	−	−	−	2.38	2.40
2	+	−	−	2.38	2.40
3	−	+	−	2.42	2.40
4	+	+	−	2.31	2.29
5	−	−	+	2.38	2.40
6	+	−	+	2.38	2.40
7	−	+	+	1.94	1.94
8	+	+	+	1.93	1.92

The main effect for Storage time = (Storage time contrast)/($r2^{k-1}$) =

$$\frac{-2.38-2.40+2.38+2.40-2.42-2.4+2.31+2.29-2.38-2.4+2.38+2.4-1.94-1.94+1.93+1.92}{2.2^{3-1}}$$

$= -0.031$

Similarly, we found that the main effect of storage temperature = -0.246, and main effect of packaging = -.211.

Similarly, for interactions; for example for the Time*Temp interaction:

The interaction effect for Time*Temp = (Time*Temp contrast)/($r2^{k-1}$) =

$$\frac{+2.38+2.40-2.38-2.40-2.42-2.4+2.31+2.29+2.38+2.4-2.38-2.4-1.94-1.94+1.93+1.92}{2.2^{3-1}}$$

$= -0.0315$

Also, the interaction effects of Storage Time and Packaging = 0.0235, and Storage temp and pack = -0.2115

(b) Using Minitab, we obtained the following ANOVA table:

```
Analysis of Variance for Color, using Adjusted SS for Tests

Source             DF    Seq SS    Adj SS    Adj MS        F      P
STime               1  0.003906  0.003906  0.003906    25.00  0.001
STemp               1  0.242556  0.242556  0.242556  1552.36  0.000
Pack                1  0.178506  0.178506  0.178506  1142.44  0.000
STime*STemp         1  0.003906  0.003906  0.003906    25.00  0.001
STime*Pack          1  0.002256  0.002256  0.002256    14.44  0.005
STemp*Pack          1  0.178506  0.178506  0.178506  1142.44  0.000
STime*STemp*Pack    1  0.002256  0.002256  0.002256    14.44  0.005
Error               8  0.001250  0.001250  0.000156
Total              15  0.613144
```

Owing to the small P-values (< 0.01), we conclude that all main effects and interactions are significant.

(c) Again using Minitab:

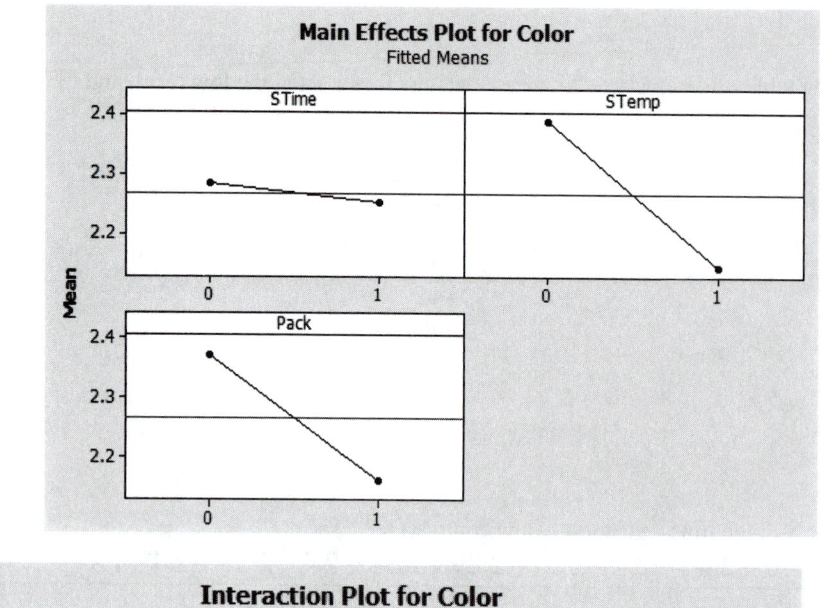

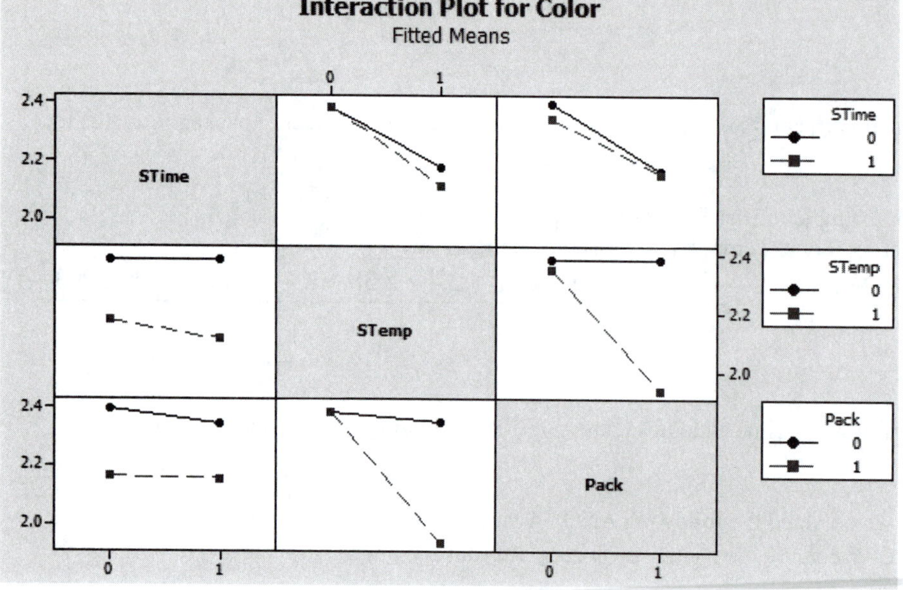

(d) All three factors (storage time, storage temperature, and packaging) should be set at their low values.

31. (a) Using the DOE option from the Minitab menus, the following effects were calculated:

```
Estimated Effects and Coefficients for combust
Term         Effect     Coef
Constant              14.312
A            -0.625   -0.313
B             9.625    4.813
C            -4.625   -2.312
A*B           0.175    0.087
A*C           1.125    0.562
B*C          -1.825   -0.913
A*B*C         2.525    1.262
```

(b) A normal probability plot can also be automatically generated within the DOE command in Minitab (by checking the 'normal plot' option). From the Minitab effects plot shown below, Factors B and C appear to be significant.

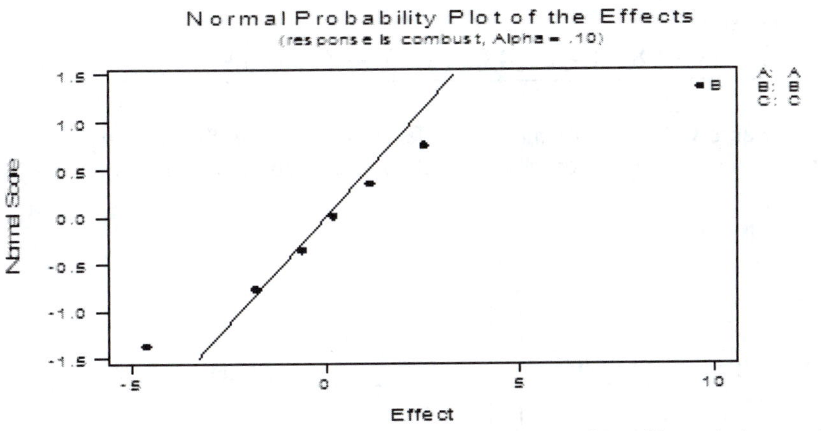

(c) Since the B effect is positive, the high level of B will maximize combustion time. Similarly, the fact that the C effect is negative means that its *low* level will maximize combustion time. Similarly, minimizing combustion time is accomplished by using the low level of B along with the high level of C.

(d) The grand average of the combustion time is 14.313. One-half of the B and C effects are 9.625/2 = 4.8125 and -4.625/2 = - 2.3125, respectively. Therefore, the prediction model is: $\hat{y} = 14.313 + 4.813x_B - 2.313x_C$

(e) For the response variable 'burnoff', Minitab gives the following effects estimates:

```
Estimated Effects and Coefficients for burnoff

Term          Effect      Coef
Constant                  4.371
A              0.002      0.001
B             -2.693     -1.346
C              0.067      0.034
A*B           -0.132     -0.066
A*C            0.108      0.054
B*C            0.063      0.031
A*B*C         -0.058     -0.029
```

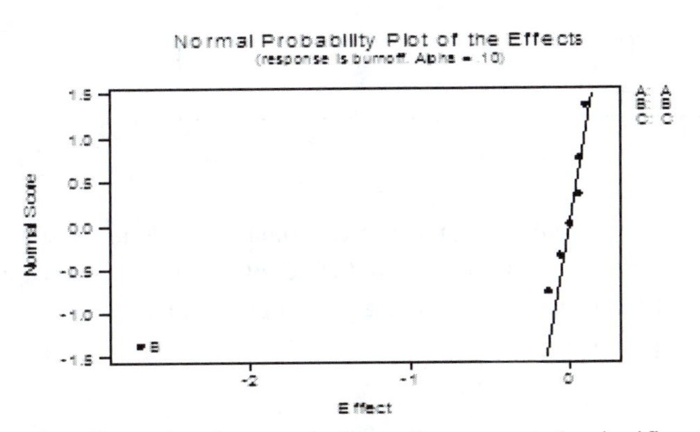

From the effects plot above, only factor B appears to be significant. Because the B effect is negative, setting B at its low level will maximize the response. Setting B at its high level will minimize the response.

The prediction equation is $\hat{y} = 4.371 - 1.347x_B$.

33. (a) Let A = time, B =current, C =EC area, D = volume, E = arsenic.

Given the table of data, where "a" represents the factor A set at the high level, and the non appearance of a denotes factor A set at the low level:
The main effect for A = (A contrast)/(r2^{k-1}) =

$$\frac{-48.70+86.50-89.1+97-58.3+84.8-\ldots-47.50+55.9-58.5+89}{2^{5-1}}=20.19$$

Notice that the coefficient of the observed value of removal (%) depends on whether the factor whose effect is being calculated was at the high level (+) or the low level (-) for that observed value of removal (%).

Similarly,

Term	Effect	Term	Effect	Term	Effect
A	20.019	AB	–3.169	BD	2.906
B	26.119	AC	–1.181	BE	–1.456
C	2.131	AD	2.131	CD	1.069
D	–17.531	AE	–1.281	CE	–0.594
E	–2.519	BC	–1.256	DE	–1.331

Note that you can also obtain these effect estimates using the DOE command in Minitab.

(b) From the probability plot of effects given below, the important effects appear to be the main effects A (time), B (current), and D (volume).

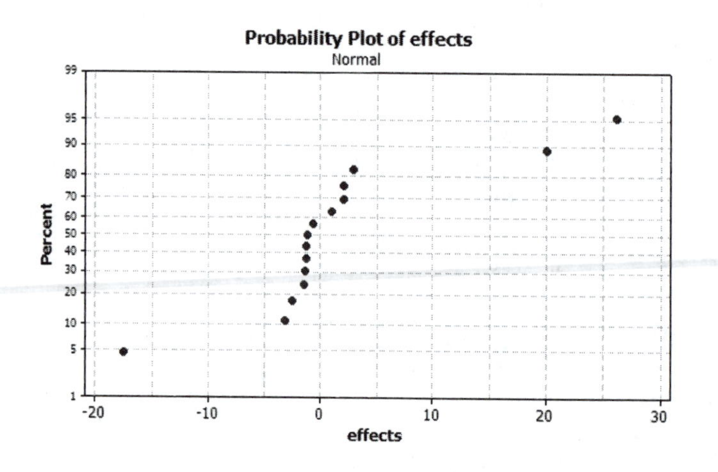

(c) A (time) and B (current) should be set to their high values. D (volume) should to set to its low value.

(d) Grand mean = 71.953, Coefficient for A = (20.019/2) = 10.010, Coefficient for B = (26.119/2) = 13.060, Coefficient for D = (–17.531/2) = –8.766. So, the prediction equation is:

$$\hat{y}=71.953+10.010x_A+13.060x_B-8.766x_D$$

Section 10.5

35.

A	B	C	D	E
-1	-1	-1	-1	1
1	-1	-1	1	-1
-1	1	-1	1	1
1	1	-1	-1	-1
-1	-1	1	1	-1
1	-1	1	-1	1
-1	1	1	-1	-1
1	1	1	1	1

37. By multiplying each of the 2^{5-1} effects through by the defining relation I = ACE = BDE = ABCD you obtain the following alias structure:

A = CE = BCD = ABDE

B = DE = ACD = ABCE

C = AE = ABD = BCDE

D = BE = ABC = ACDE

E = AC = BD = ABCDE

AB = CD = ADE = BCE

AD = BC = ABE = CDE

The entries in this alias structure can be verified by multiplication.

39. (a) k = 5; p = 1

(b) Alias structure

I = ABCDE	AD = BCE
A = BCDE	AE = BCD
B = ACDE	BC = ADE
C = ABDE	BD = ACE
D = ABCE	BE = ACD
E = ABCD	CD = ABE
AB = CDE	CE = ABD
AC = BDE	DE = ABC

(c) No. The four–way interactions are confounded with the main effects. The two–way interactions are confounded with the three–way interactions. So, if all interactions consisting of three or more factors are negligible, none of the estimates of the remaining effects will be confounded with one another.

41. (a) Let factor A = Board orientation, B = Anode height, and C = Anode placements. This is a 2^{3-1} fractional factorial design. (i.e., 4 runs) So, k = 3 and p = 1.

(b) The design generator in this design is C = -AB. The alias structure is:

A = - BC
B = - AC
C = - AB

(c)

Effect Name	Effect
A	-3.135
B	-1.135
C	-4.925

(d) Assuming the AB interaction is negligible is equivalent to assuming the factor C effect is negligible. [Note: Assuming a main effect is negligible is probably <u>not</u> a good idea.] However, the sums of squares required for the F-tests are as follows:

$SSE = (-4.925)^2 = 24.26$
$SSTo = (s^2)(3) = (3.4338)^2(3) = 35.37$
$SSA = (-3.135)^2 = 9.83$
$SSB = (-1.135)^2 = 1.29$

The corresponding ANOVA table produced in Minitab is:

Analysis of Variance for variation in thickness

Source	df	SS	MS	F	P-value
A	1	9.83	9.83	.41	.639
B	1	1.29	1.29	.05	.856
Error	1	24.26	24.26		
Total	3	35.37			

When testing at $\alpha = .05$, we find that neither factor A nor factor B is important, since their corresponding P-values (.639 and .856) are so large.

(e) Based on our analysis in part (d), we cannot conclude that factors A or B are significant. Also, we assumed factor C was not significant in order to test for the significance of factors A and B.

(f) Since we have found no significant differences between the factors, the decision about how to minimize the variation in plating thickness would not be made using the statistical analysis from early parts of this problem. However, based solely on the sign of the effect for each factor, one might conclude that all three factors should be set at the high level (+1), in order to minimize the variation in plating thickness.

43. (a) Let A = temp, B = pH, C = yeast, D = Tryptone, and E = Nitsch. Using the DOE command in Minitab, the effects estimates are:

Term	Effect	Coef	Term	Effect	Coef
Constant		50.288	*AD*	–1.400	–0.700
A	23.750	11.875	*AE*	–1.050	–0.525
B	6.850	3.425	*BC*	12.450	6.225
C	–0.675	–0.337	*BD*	14.100	7.050
D	–18.725	–9.363	*BE*	–2.700	–1.350
E	–5.725	–2.863	*CD*	–4.125	–2.062
AB	–8.075	–4.037	*CE*	8.225	4.112
AC	4.200	2.100	*DE*	–6.675	–3.337

Estimates of the 3 and 4–way interaction terms would be determined by the estimates of the corresponding aliased term. See Exercise 39 for alias structure.

(b) A (= $BCDE$), D (=$ABCE$)

(c) Here are the main effects plots for factors A and D

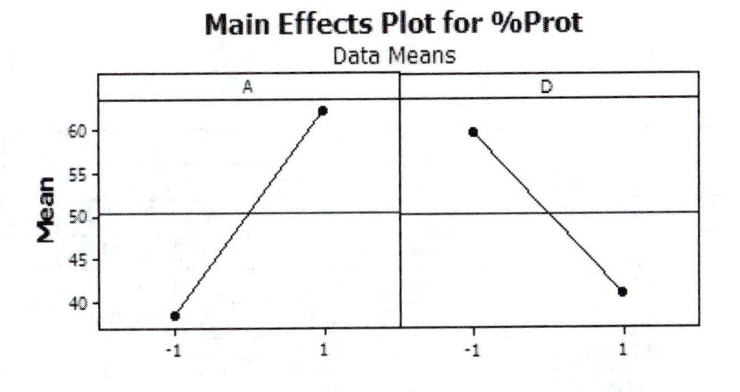

(d) Settings that maximize percent protection are: A high and D low.

Supplementary Exercises

45. (a) This is a two-factor experiment where factor A has 3 levels and factor B has 4 levels. The 12 factor-level combinations were each replicated 3 times producing 36 experimental runs.

The ANOVA table for this experiment is:

```
Analysis of Variance for compressive strength
```

Source	df	SS	MS	F	$F_{0.05}$
A	2	30,763.00	15,381.50	3.79	3.40
B	3	34,185.60	11,395.20	2.81	3.01
AB	6	43,581.20	7,263.53	1.79	2.51
Error	24	97,436.80	4,059.87		
Total	35	205,966.60			

Critical values corresponding to each factor were identified using Table VIII at $\alpha = .05$.

(b) H_O: There is no AB interaction.

Since 1.79 $< F_{.05}(6,24) = 2.51$, we fail to reject H_O. There is not sufficient evidence that a significant interaction exists between factors A and B.

(c) H_o: There is no main effect for factor A.

Since $3.79 > F_{.05}(2, 24) = 3.40$, we reject H_o. There is sufficient evidence to claim that factor A affects compressive strength.

(d) H_O: There is no main effect for factor B.

Since $2.81 < F_{.05}(3,24) = 3.01$, we fail to reject H_O. There is not sufficient evidence to claim that factor B affects compressive strength.

47. (a) We know that:

df for A = 3-1 = 2; df for B = 3-1 = 2; df for C = 3-1 = 2;
df for AB = (3-1)*(3-1) = 4; df for AC = (3-1)*(3-1) = 4; df for BC = (3-1)*(3-1) = 4;
Total df = 3*3*3 - 1 = 27-1 = 26
Error df = 26 – (2+2+2+4+4+4) = 8

And any MS = SS/df; $F_{factor} = MS_{factor}/MSE$

The ANOVA table

Source	df	SS	MS	F	P-value
A	2	2.0742421	1.03712106	162.38	< 0.001
B	2	0.08057	0.040285	6.30734304	0.023
C	2	0.26039	0.130195	20.3843745	<0.001
AB	4	0.0143069	0.00357672	0.56	0.698
AC	4	0.145137	0.03628425	5.6809535	0.018
BC	4	0.0194165	0.00485412	0.76	0.579
Error	8	0.051096	0.006387		
Total	26				

(b) The main effect of laser power (A) and main effect of powder flow rate (C) are significant, because both P-values corresponding to the calculated F-statistics, with $df_1 = 2$ and $df_2 = 8$ turn out to be < 0.001 which is less than 0.01. The remaining main effect of B, and the two-factor interactions are not significant.

49. (a) Using Minitab, we found the following ANOVA table:

```
Analysis of Variance for Ni, using Adjusted SS for Tests

Source             DF  Seq SS  Adj SS  Adj MS     F      P
Power               2  326.67  326.67  163.34  5.89  0.027
Speed               2   43.83   43.83   21.92  0.79  0.486
thickness           2  123.84  123.84   61.92  2.23  0.169
Power*Speed         4   48.51   48.51   12.13  0.44  0.779
Power*thickness     4  168.26  168.26   42.06  1.52  0.285
Speed*thickness     4   23.49   23.49    5.87  0.21  0.925
Error               8  221.68  221.68   27.71
Total              26  956.28
```

(b) From the ANOVA table (see above), we can see that none of the two-factor interactions have a P-value that is less than 0.05. Thus, there is no evidence that any of the two-factor interaction effects are significant.

(c) From the ANOVA table, we can see that the main effect of power is significant because the P-value = 0.027 < 0.05.

51. (a) Using the DOE command in Minitab, we can obtain the following effect estimates and coefficients

Factorial Fit: Lignin versus Time, Press, Temp

```
Estimated Effects and Coefficients for Lignin (coded units)

Term               Effect     Coef
Constant                    163.88
Time                19.25     9.63
Press               99.25    49.62
Temp                41.25    20.63
Time*Press         -63.25   -31.62
Time*Temp            3.75     1.88
Press*Temp         -47.25   -23.62
Time*Press*Temp      1.25     0.62
```

(b) Here's a probability plot of the effects:

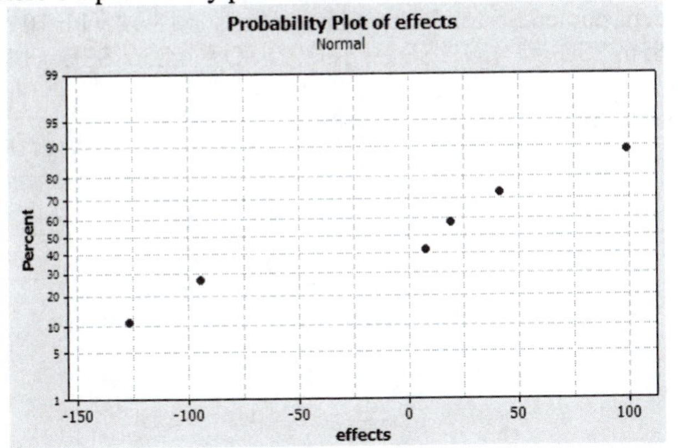

(c) From the probability plot above, it appears that the significant effects are: Pressure, Temperature, Time*Pressure, and Pressure*Temp.

(d) Here are the main effects plots for Pressure and Temperature:

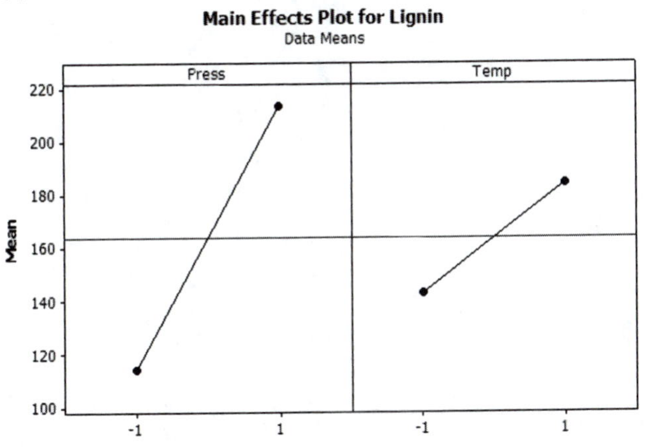

Here are the interaction plots for Time*Pressure and Pressure*Temp

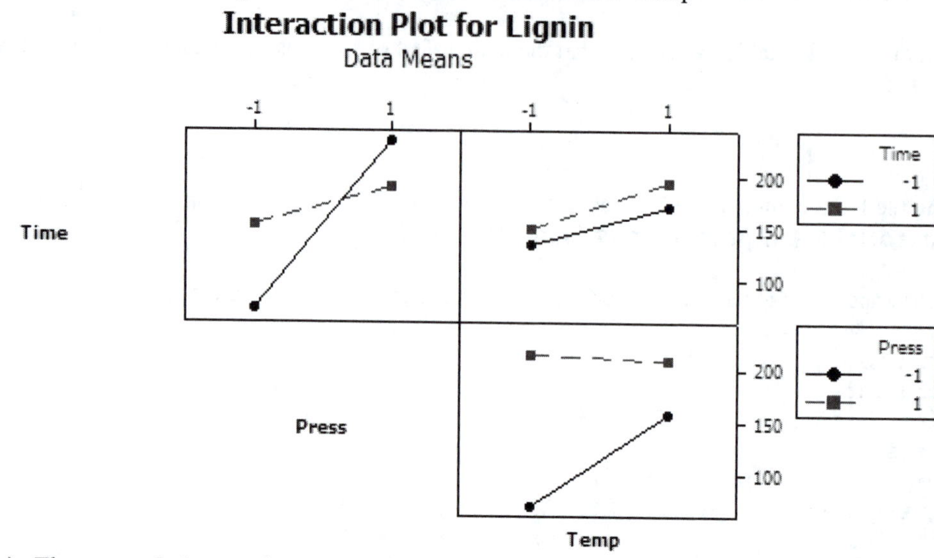

(e) Then we only have to be concerned with the settings of the factors Pressure (B) and Temperature (C). Set both Pressure and Temperature at their high settings.

53. Caution: test runs are not in Yates order; pooled SS for 2-factor interactions is 18.12 with 10 degrees of freedom, so MSE = 1.812; SSA = 0.856, SSB = 11.391, SSC = 1.380, SSD = 44.56, SSE = 14.25. Factors B, D, and E are significant at $\alpha = .01$.

Chapter 11

Inferential Methods in Regression and Correlation

Section 11.1

1. (a) The slope of the estimated regression line ($\beta = .095$) is the expected <u>change</u> of in the response variable y for each one-unit <u>increase</u> in the x variable. This, of course, is just the usual interpretation of the slope of a straight line. Since x is measured in inches, a one-unit increase in x corresponds to a one-inch increase in pressure drop. Therefore, the expected change in flow rate is .095 m^3/min.

   (b) When the pressure drop, x, changes from 10 inches to 15 inches, then a 5 unit increase in x has occurred. Therefore, using the definition of the slope from (a), we expect about a 5(.095) = .475 m^3/min. increase in flow rate (it is an *increase* since the sign of $\beta = .095$ is *positive*).

   (c) For x = 10, $\mu_{y.10} = -.12 + .095(10) = .830$. For x = 15, $\mu_{y.15} = -.12 + .095(15) = 1.305$.

   (d) When x = 10, the flow rate y is normally distributed with a mean value of $\mu_{y.10} = .830$ and a standard deviation of $\sigma_{y.10} = \sigma = .025$. Therefore, we standardize and use the z table to find: $P(y > .835) = P(z > \frac{.835-.830}{.025}) = P(z > .20) = 1 - P(z \le .20) = 1 - .5793 = .4207$ (using Table I).

3. (a) The simple linear regression model states that $y = \alpha + \beta x$. Here, $y = \ln(V)$ and $x = 1/T$, so the model becomes $\ln(V) = \alpha + \beta(1/T) + e$. Exponentiating both sides gives: $\exp(\ln(V)) = \exp(\alpha + \beta(1/T)+e)$, or, $V = e^{\alpha}e^{\beta/T}\varepsilon = e^{\alpha}(e^{\beta})^{1/T}\varepsilon = \gamma_0(\gamma_1)^{1/T}\varepsilon$, where ε is the antilog of the error term e, $\gamma_0 = e^{\alpha}$, and $\gamma_1 = e^{\beta}$. To summarize, the model is $V = \gamma_0(\gamma_1)^{1/T}\varepsilon$, which is the model for a multiplicative relationship between the response and predictor variables.

   (b) For estimation, we normally set the error term equal to its expected value. For a <u>multiplicative</u> model, the expected value of the error term is 1 (for additive models the expected value is 0). So, for $\alpha = 20.607$, $\beta = -5200.762$ and a temperature of T = 300, we predict a vapor pressure of about $V = \gamma_0(\gamma_1)^{1/T}\varepsilon = (e^{20.607})(e^{-5200.762})^{1/300}(1) = 26.341$.

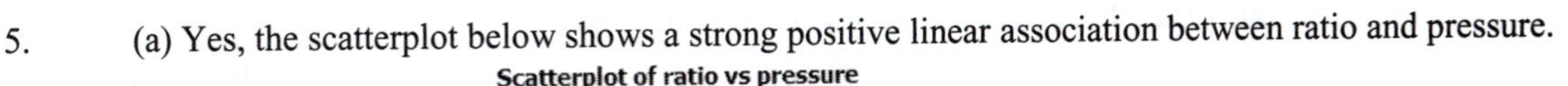

5. (a) Yes, the scatterplot below shows a strong positive linear association between ratio and pressure.

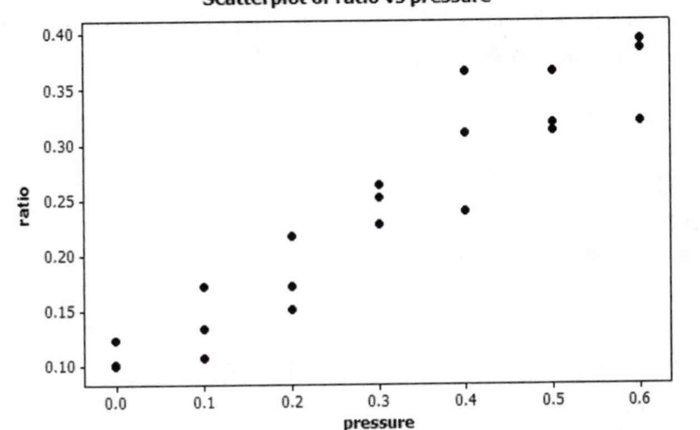

   (b) Using Minitab, we can obtain the following simple linear regression output:

Regression Analysis: ratio versus pressure

```
The regression equation is
ratio = 0.101 + 0.461 pressure

S = 0.0332397   R-Sq = 89.5%   R-Sq(adj) = 88.9%
```

Hence, the estimated slope = 0.461; estimated intercept = 0.101

(c) Let's plug in 0.45 for pressure in the equation. Then,

Predicted ratio = 0.101 + 0.461(0.45) = 0.30854

(d) From the Minitab output, Estimated error SD = 0.033

7. (a) Yes, the relationships look fairly linear, as can be seen in the scatterplots below:

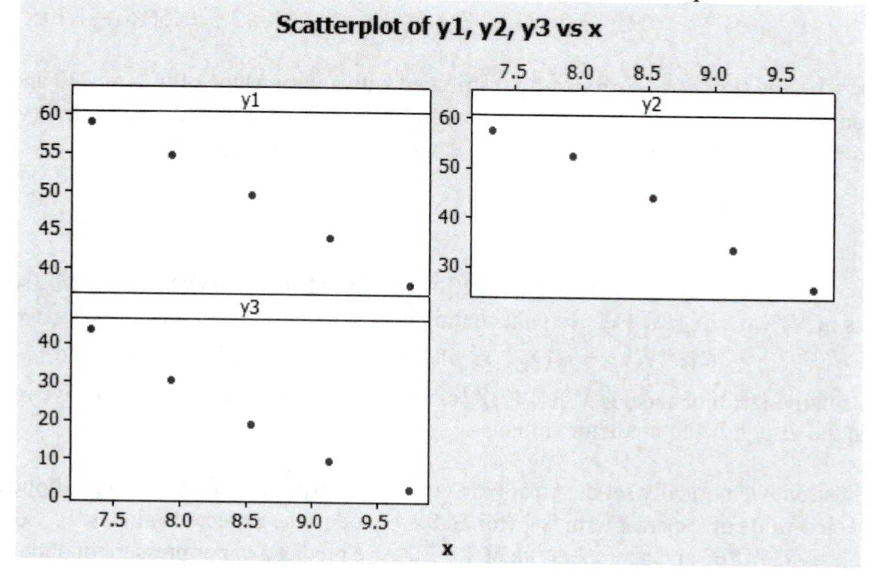

(b) Using Minitab, we can obtain the following estimates for slope and intercept:

For (x, y'): Estimated slope = -8.71; estimated intercept =123
For (x, y''): Estimated slope = -13.6; estimated intercept = 159
For (x, y'''): Estimated slope = -17.1; estimated intercept = 167

(c) As timber damage increases from 0% to 20% to 40%, the decrease in critical rating associated with every 1m increase in pile length gets bigger (from 8.71, to 13.6, to 17.1).

(d) Also from Minitab, we can find that:

For (x, y'): Estimated error SD = 0.668
For (x, y''): Estimated error SD = 1.734
For (x, y'''): Estimated error SD = 2.156

Yes, the error SDs tend to increase as timber damage increases from 0% to 20% to 40%.

9. (a) For a one unit increase in inverse foil thickness, one would expect a .260 unit increase in flux.

Secondly, 98% of the observed variation in flux can be attributed to the simple linear regression relationship between flux and inverse foil thickness.

(b) $\mu_{y \bullet 23.5} = -.398 + .260(23.5) = 5.712$

(c) A value of flux that would result from a single observation made when inverse foil thickness is 45 is:

-.398 + .260(45) = 11.302

(d) The sum of the residuals provided by Minitab do, in fact, sum to zero.

That is: (-.458 + .634 + … - .011) = 0

Secondly,

$$\Sigma(\text{residuals})^2 = \Sigma\left((-.458)^2 + (.634)^2 + ... + (-.011)^2\right) = 1.219$$

This value is close to the SSResid value given in the Minitab output (1.218).

Section 11.2

11. H_0: $\alpha = 0$ versus H_a: $\alpha \neq 0$.

The calculated t statistic is:

$$t = \left(\frac{a}{s_a}\right) = \left(\frac{-1.128}{2.368}\right) = -.48$$

The corresponding P-value = .642. Therefore, since the P-value is large, we would not reject the null hypothesis. We cannot conclude that the vertical intercept of the population line is nonzero.

13. (a) **Method 1:** Hypothesis Test, H_0: $\beta = 0$ versus H_a: $\beta \neq 0$

Where β = change in average amount of oil recovered associated with every one unit increase in amount of oil added.

Using Minitab, we can find the following:

Regression Analysis: Recovered versus Added

```
The regression equation is Recovered = - 0.523 + 0.878 Added

Predictor      Coef   SE Coef      T      P
Constant    -0.5234    0.1453  -3.60  0.003
Added       0.87825   0.01610  54.56  0.000
```

Notice that t = 54.56, and P-value < .0001, reject H_0 and conclude that there is a useful linear relationship between these two variables.

Method 2: A confidence interval for $\beta = b \pm (t \text{ critical value}) \cdot s_b$.

A 95% confidence interval for β is: .87825 ± (2.179)(.01610) = (0.8432, 0.9133), using t critical value for df = $(n - 2)$ = (14 – 2) = 12. The plausible values are all positive so we conclude there is a useful linear relationship between the two variables.

(b) The t ratio for testing model utility would be the same value regardless of which of the two variables was defined to be the independent variable. This can be easily seen by looking at the t test statistic for testing if the population correlation coefficient is equal to zero. In that equation the only values required are the sample size (n) and the sample correlation coefficient (r). Both r and n are not dependent on which variable was the independent variable.

15. The t ratio for testing the model utility is dependent only on the sample size and the sample correlation coefficient. Neither of these quantities is unit dependent. So, multiplying the dependent variable by a constant will have no effect on the t test statistic.

17. H_0: $\rho = 0$ versus H_a: $\rho > 0$.

The test statistic is: $t = \left[\dfrac{r\sqrt{n-2}}{\sqrt{1-r^2}} \right]$

So, we need: $r = \left[\dfrac{.2073}{\sqrt{1.183\sqrt{.05080}}} \right] = .8456$

Therefore, $t = \left[\dfrac{(.8456)\sqrt{13-2}}{\sqrt{1-(.8456)^2}} \right] = 5.25$

The corresponding P-value = $P(t > 5.25)$. With df = $(n - 2) = 11$ and using Minitab to obtain the precise P-value, we obtain a P-value of $< .0001$.

With such a small P-value we would reject the null hypothesis and conclude that there is a positive linear relationship between these two variables.

19. (a) A simple linear regression model is given by $\hat{y} = a + bx$, where $b = S_{xy} / S_{xx}$.

$S_{xy} = \sum x_i y_i - (1/n)\left(\sum x_i\right)\left(\sum y_i\right) = 2759.6 - (1/19)(221.1)(193) = 249.4647$

$S_{xx} = \sum x_i^2 - (1/n)\left(\sum x_i\right)^2 = 3056.69 - (1/17)(221.1)^2 = 181.0894$

So $b = S_{xy} / S_{xx} = 249.4647 / 181.0894 = 1.37758$

We can compute a 95% confidence interval for the true population slope coefficient β:
We compute $b \pm s_b \cdot t$, where t is the appropriate critical value with $n - 2 = 17 - 2 = 15$ df. So the appropriate critical t-value for a 95% confidence interval is 2.131.

$s_b = s_e / \sqrt{S_{xx}}$, where $s_e = \sqrt{\text{SSResid}/(n-2)}$, and $\text{SSResid} = \text{SSTo} - b \cdot S_{xy}$

$\text{SSTo} = S_{yy} = \sum y_i^2 - (1/n)\left(\sum y_i\right)^2 = 2975 - (1/17)(193)^2 = 783.88235$

$\text{SSResid} = \text{SSTo} - b \cdot S_{xy} = 783.88235 - (1.37758)(249.4647) = 440.22$

$s_e = \sqrt{\text{SSResid}/(n-2)} = \sqrt{440.22/(17-2)} = 5.417$

$s_b = s_e / \sqrt{S_{xx}} = 5.417 / \sqrt{181.0894} \approx .4025$

So a 95% confidence interval for β is given by $1.37758 \pm (.4025)(2.131)$, which gives the interval (.51985, 2.2531). Therefore, we are 95% confident that the true mean percentage increase in nausea is between .51985% and 2.23531% for a 1 unit increase in dose.

(b) Yes, there does appear to be a meaningful linear relationship between the two variables. Our 95% confidence interval for β is entirely positive, suggesting that a positive, linear relationship exits.

(c) No, we should not try to predict y at the x value of 5.0, since it is smaller than the smallest x-value used in the regression analysis. Remember, we should not extrapolate outside the range of x-values used for the regression analysis.

(d) By deleting the observation $(x, y) = (6.0, 2.50)$, our new summary statistics become:

$n = 17 - 1 = 16;\ \sum x_i = 221.1 - 6.0 = 215.1;\ \sum y_i = 193 - 2.50 = 190.50;$

$\sum x_i^2 = 3056.69 - 6.0^2 = 3020.69;\ \sum x_i y_i = 2759.6 - (6.0)(2.50) = 2744.6;$

$\sum y_i^2 = 2975 - 2.50^2 = 2968.75$

$$S_{xy} = \sum x_i y_i - (1/n)\left(\sum x_i\right)\left(\sum y_i\right) = 2744.6 - (1/16)(215.1)(190.50) = 183.5656;$$

$$S_{xx} = \sum x_i^2 - (1/n)\left(\sum x_i\right)^2 = 3020.69 - (1/16)(215.1)^2 = 128.9394$$

So $b = S_{xy} / S_{xx} = 183.5656 / 128.9894 \approx 1.42366$

By deleting $(x, y) = (6.0, 2.50)$, our new estimate for β is $b = 1.42366$, which is still fairly close to the value of $b = 1.37758$ that was computed in part (a) above. Moreover, this new estimate of $b = 1.42366$ falls well within the 95% CI that was obtained in part (a) above. Therefore, the point (6.0, 2.50) does not appear to exert any undue influence on our regression analysis.

Section 11.3

21. (a) The following scatter plot appears to be quite linear. So, yes, a scatter plot does support the use of a simple linear regression analysis.

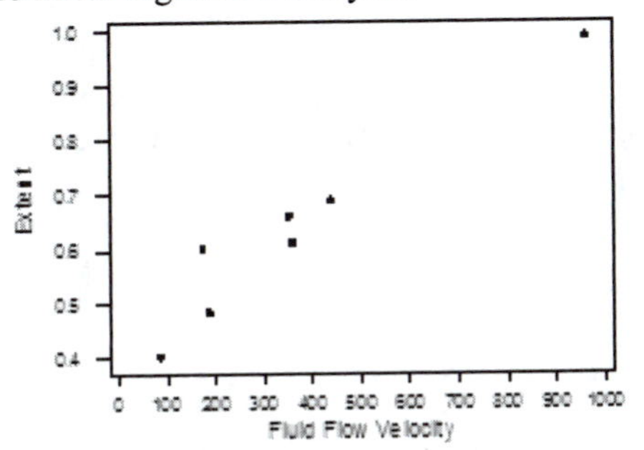

(b) We desire to compute r^2, which is given by the formula $r^2 = 1 - \text{SSResid}/\text{SSTo}$.

$$\text{SSTo} = S_{yy} = \sum y_i^2 - (1/n)\left(\sum y_i\right)^2 = 3.0143 - (1/7)(4.43)^2 = .21074$$

$$\text{SSResid} = \text{SSTo} - b \cdot S_{xy}$$

$$b = S_{xy} / S_{xx};$$

$$S_{xy} = \sum x_i y_i - (1/n)\left(\sum x_i\right)\left(\sum y_i\right) = 1947.31 - (1/7)(2578)(4.43) = 315.8042857$$

$$S_{xx} = \sum x_i^2 - (1/n)\left(\sum x_i\right)^2 = 1{,}457{,}920 - (1/7)(2578)^2 = 508{,}479.4286$$

So $b = S_{xy} / S_{xx} = 315.804/508{,}479 \approx 6.2108 \times 10^{-4}$

Therefore, $\text{SSResid} = \text{SSTo} - b \cdot S_{xy} = .21074 - (6.2108 \times 10^{-4})(315.8042857) \approx .0146$

Thus, $r^2 = 1 - \text{SSResid}/\text{SSTo} = 1 - (.0146/.21074) = .931$.

So 93.1% of the observed variation in mist can be attributed to a simple linear relationship between velocity and mist.

(c) If increasing velocity by 900 cm/sec results in an average change in the response of .6, then our true population slope coefficient is $\beta = .6/900 = 6.667 \times 10^{-4}$.

We now desire to test the hypothesis $H_0 : \beta = 6.667 \times 10^{-4}$ versus $H_a : \beta < 6.667 \times 10^{-4}$. The test-statistic is given by $t = (t - \beta)/s_b$, which follows a *t*-distribution with $n - 2 = 7 - 2 = 5$ degrees of freedom. (This is the appropriate test-statistic for simple linear regression.) This is a one-sided, left-tailed test. At the significance level of $\alpha = .05$, we use the critical value of –2.015.

We now require $s_b = s_e / \sqrt{S_{xx}}$, where $s_e = \sqrt{\text{SSResid}/(n-2)} = \sqrt{.0146/(7-2)} = .054037$.
Hence, $s_b = s_e / \sqrt{S_{xx}} = .054037 / \sqrt{508{,}479.4286} = 7.578 \times 10^{-5}$.

Our test-statistic is $t = \dfrac{b - \beta}{s_b} = \dfrac{6.2108 \times 10^{-4} - 6.667 \times 10^{-4}}{7.578 \times 10^{-5}} = -.6016$, which is greater than the critical value of –2.015. Thus, we fail to reject H_0, and conclude that if *x* increases by 900 units, the true average increase in *y* is not substantially less than .6.

(d) We will estimate β using the 95% confidence interval given by $b \pm s_b \cdot t$:
$6.21075 \times 10^{-4} \pm (7.578 \times 10^{-5})(2.571) \Rightarrow (4.26 \times 10^{-4}, 8.159 \times 10^{-4})$
We are 95% confident that the true average change in mist associated with a 1 cm/sec increase in velocity is between 4.26×10^{-4} and 8.159×10^{-4}.

23. (a) 1500 degrees Fahrenheit will be expressed as 1.5 in our regression equation. In order to compute a 95% prediction interval we need the following quantities, most of which were obtained from exercise 5.

$\hat{y} = -2.0951 + 3.6933(1.5) = 3.445$

$s_e = .0644$

$$s_{\hat{y}} = \left[.0644\sqrt{\frac{1}{9} + \frac{(1.5 - 1.4)^2}{.6}}\right] = .02302$$

So, a 95% prediction interval is: (note: df = 7)

$$3.445 \pm (2.365)\sqrt{(.0644)^2 + (.02302)^2}$$
$$3.445 \pm .1617$$
$$(3.2833 \text{ , } 3.6067)$$

Therefore, we would predict the oxygen diffusivity for a single observation to be made when the temperature is 1500 degrees to be between 3.28 and 3.61.

(b) The interval when the temperature is 1200 degrees will be wider than when the temperature is 1500 degrees. This is because 1200 degrees is 200 degrees away from the mean temperature of 1400 degrees whereas 1500 degrees is only 100 degrees away from the mean temperature.

25. The mean x value is 40.3. Intervals with x values farther away from this mean are wider.

Also, prediction intervals are wider than confidence intervals. And, 99% intervals are wider than 95% intervals.

Therefore,

i will be wider than iii
i will be more narrow than ii
ii will be wider than iv
iii will be more narrow than iv and v

27. (a) Let β denote the true average change in milk protein for each 1 kg/day increase in milk production. The relevant hypotheses to test are $H_0: \beta = 0$ versus $H_a: \beta \neq 0$. The test statistic is $t = b/s_b$ based on n-2 = 14-2 = 12 degrees of freedom. In order to find s_b, we first find s_e:
$s_e^2 = \frac{SS\,Resid}{n-2} = \frac{.02120}{12} = .001767$, so $s_e = .0420$.

Then, $s_b = \frac{s_e}{\sqrt{SS_{xx}}} = \frac{.0420}{\sqrt{762.012}} = .00152$, which then gives a calculated t value of $t = b/s_b$ = $.024576/.00152 \approx 16.2$. The 2-sided P-value associated with t = 16.2 is approximately 2(.000) = .000 (Table VI), so H_0 is rejected in favor of the conclusion that there is a useful linear relationship between protein and production. We should not be surprised by this result since the r^2 value for this data is .956.

(b) For a 99% confidence interval based on d.f. = 12, the t-critical value is 3.055 (from Table IV). The estimated regression line gives a value of $\hat{y} = .175576 + .024576(30) = .913$ when x = 30. Therefore, $s_{\hat{y}} = (.0420)\sqrt{\frac{1}{14} + \frac{(30-29.56)^2}{762.012}} = .01124$ and the 95% confidence interval is then: .913 $\pm(3.055)(.01124) = .913 \pm .034$ = [.879, .947].

(c) The 99% prediction interval for protein from a single cow is:
$\hat{y} \pm (\text{t-critical})\sqrt{s_e^2 + s_{\hat{y}}^2} = .913 \pm (3.055)\sqrt{(.0420)^2 + (.01124)^2}$
$= .913 \pm .133 = [.780, 1.046]$.

Section 11.4

29. (a) The mean value of y, when $x_1 = 50$ and $x_2 = 3$ is $-.800 + .060(50) + .900(3) = 4.9$ hours.

(b) When the number of deliveries (x_2) is held fixed, then average change in travel time associated with a one-mile (i.e., one unit) increase in distance traveled (x_1) is .060 hours. Similarly, when the distance traveled (x_1) is held fixed, then the average change in travel time associated with one extra delivery (i.e., a one-unit increase in x_2) is .900 hours.

(c) Under the assumption that y follows a normal distribution, the mean and standard deviation of this distribution are 4.9 (because $x_1 = 50$ and $x_2 = 3$) and $\sigma = .5$ (since σ is assumed to be constant regardless of the values of x_1 and x_2). Therefore, $P(y \leq 6) = P(z \leq (6\text{-}4.9)/.5) = P(z \leq 2.20) = .9861$ (from Table I). That is, in the long run, about 98.6% of all days will result in a travel time of at most 6 hours.

31. (a) A graph created using Minitab is given below:

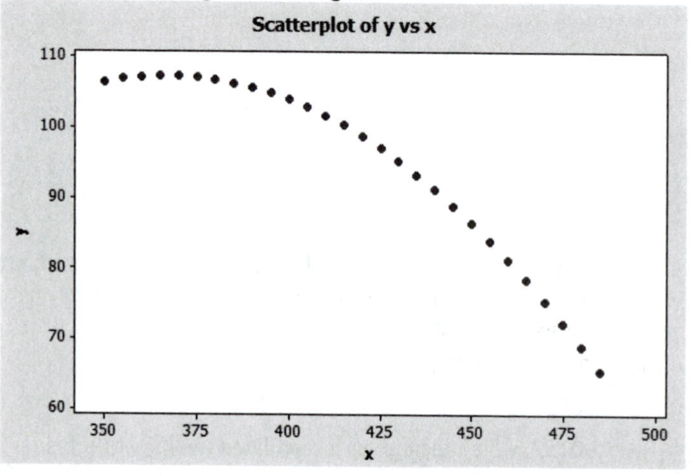

(b) The mean free flow percentage (y) would be higher for a viscosity (x) value of 450, compared to a viscosity value of 470, as can be seen from the graph.

(c) When viscosity, x = 450 ➔ mean free flow percentage, y = 86.5
When viscosity, x = 460 ➔ mean free flow percentage, y = 81.2
Thus the change in mean free flow percentage when viscosity goes from 450 to 460, is a decrease of 5.3

When viscosity, x = 470 ➔ mean free flow percentage, y = 75.3
Thus the change in mean free flow percentage when viscosity goes from 460 to 470, is a decrease of 5.9

33. (a) For $x_1 = 2$, $x_2 = 8$ (remember, the units of x_2 are in 1000's), and $x_3 = 1$ (since the outlet has a drive-up window), the average sales are $y = 10.00 - 1.2(2) + 6.8(8) + 15.3(1) = 77.3$ (i.e, \$77,300).

(b) For $x_1 = 3$, $x_2 = 5$, and $x_3 = 0$, the average sales are $y = 10.00 - 1.2(3) + 6.8(5) + 15.3(0) = 40.4$ (i.e, \$40,400).

Section 11.5

35. (a) The appropriate hypotheses are $H_0: \beta_1 = \beta_2 = 0$ versus H_a: *at least one of the β_i's is not zero.* The test statistic is F = 1260.71, and the corresponding P-value is < 0.001. Therefore, H_0 can be rejected at any reasonable level of significance. We conclude that at least one of the two predictor variables appears to provide useful information about deposition rate of a welding process.

(b) A 95% confidence interval for β_2: $b_2 \pm$ (t critical value) s_{b2} , where df = 19.

Thus the 95% confidence interval: 0.002775 ± (2.093)(.001121) = (0.00043, 0.00512)

(c) By plugging in, we can find that $\hat{y} = 4.478$. Then, the 95% confidence interval for the true average deposition rate is: 4.478 ± (2.093).(.02438) = (4.427, 4.529)

(d) The 95% prediction interval is given by:

$$4.478 \pm (2.093).\sqrt{.044853^2 + .02438^2} = (4.371, 4.585)$$

37. (a) The appropriate hypotheses are $H_0: \beta_1 = \beta_2 = \beta_3 = \beta_4 = 0$ versus H_a: *at least one of the β_i's is not zero.*

The test statistic is $F = \dfrac{R^2/k}{(1-R^2)/(n-(k+1))} = \dfrac{.946/4}{(1-.946)/(25-(4+1))} = 87.6$. The test is based on df_1 = 4, df_2 = 20. From Table XII, the P-value associated with F = 6.59 is .001, so the P-value associated with 87.6 is obviously .000. Therefore, H_0 can be rejected at any reasonable level of significance. We conclude that at least one of the four predictor variables appears to provide useful information about tenacity.

(b) The adjusted R^2 value is $1 - \dfrac{n-1}{n-(k+1)} \dfrac{SSResid}{SSTo} = = 1 - \dfrac{n-1}{n-(k+1)}[1-R^2] = 1 - \dfrac{24}{20}[1-.946] =$ .935, which does not differ much from R^2 = .946.

(c) The estimated average tenacity when x_1 = 16.5, x_2 = 50, x_3 = 3, and x_4 = 5 is: $\hat{y} = 6.121 - .082x_1 + .113x_2 + .256x_3 - .219x_4$ = 6.121 -.082(16.5) +.113(50) +.256(3) -.219(5) = 10.091. For a 99% confidence interval based on 20 d.f., the t-critical value is 2.845. The desired interval is: 10.091 ± (2.845)(.350) = 10.091 ± .996, or, about [9.095, 11.087]. Therefore, when the four predictors are as specified in this problem, the true average tenacity is estimated to be between 9.095 and 11.087.

39. (a) The negative value of b_2, which is the coefficient of x^2 in the model, indicates that the parabola $b_0 + b_1x + b_2x^2$ opens downward.

(b) R^2 = 1 - SSResid/Ssto = 1 - .29/202.87 = .9986, so about 99.86% of the variation in output power can be attributed to the relationship between power and frequency.

(c) With an R^2 this high, it is very likely that the test statistic will be significant. The relevant hypotheses are $H_0: \beta_1 = \beta_2 = 0$ versus H_a: *at least one of the β_i's is not zero.* The test statistic is: F = $\dfrac{SSRegr/k}{SSResid/(n-(k+1))} = \dfrac{(202.87-2.9)/2}{.29/5} = 1746$. Clearly, the P-value associated with F = 1746 is 0, so H_0 is rejected and we conclude that the model is useful for predicting power.

(d) The relevant hypotheses are $H_0: \beta_2 = 0$ versus $H_a: \beta_2 \neq 0$. The test statistic is: $t = b_2 / s_{b_2}$ = -.00163141/.00003391 = 48. The P-value for this statistic is 0 and H_0 is rejected in favor of the conclusion that the quadratic predictor provides useful information.

(e) The estimated average power when x = 150 is $\hat{y}$ = -1.5127 + .391902x - .00163141x^2 = -1.5127 + .391902(150) - .00163141$(150)^2$ = 20.57. The t-critical value based on 5d.f. is 4.032, so the 99% confidence interval is: 20.57 ± (4.032)(.1410) = 20.57 ± .57 or, about [20.00, 21.14] To find the prediction interval, we must first find s_e. s_e^2 = SSResid/(n-3) = .29/5 = .058, so s_e = .241. Therefore, the 99% prediction interval is:

$$20.57 \pm (4.032)\sqrt{(.241)^2 + (.141)^2} = 20.57 \pm 1.13, \text{ or, about } [19.44, 21.70].$$

41. Let β_1, β_2, β_3, β_4, and β_5 denote the regression coefficients for x_1, x_2, x_1^2, x_2^2, and x_1x_2, respectively. We then wish to test $H_0: \beta_3 = \beta_4 = \beta_5 = 0$ versus H_a: *at least one of β_3, β_4, and β_5 is not zero.* H_0 asserts that none of the three second-order variables provides useful information. The test statistic is:

$$F = \frac{[\text{SSResid(reduced)} - \text{SSResid(full)}]/g}{\text{SSResid(full)}/[n - (k+1)]} = \frac{[894.95 - 390.64]/3}{390.64/[14 - (5+1)]} = 3.44,$$

where g = the number of predictors in the group considered for deletion = 3, k = the number of predictors in the full model = 5, and n = the sample size = 14. The test is based on $df_1 = 3$ and $df_2 = 8$. From the Table XII we find .05 < P-value < .10. Therefore, at significance level $\alpha = .01$, H_0 cannot be rejected. That is, we do not have evidence that any of the three second-order predictors provide useful information beyond what is already provided by x_1 and x_2 together. Although there seems to be a large difference between the SSResid values for the full and reduced models, the small number of degrees of freedom for error (d.f. error = 8) is simply too small to allow us to conclude that the full model is useful.

43. (a) The variable "supplier" has three categories, so we need two indicator variables to code "supplier," such as

$$x_2 = \begin{cases} 1 & \text{supplier 1} \\ 0 & \text{otherwise} \end{cases} \qquad x_3 = \begin{cases} 1 & \text{supplier 2} \\ 0 & \text{otherwise} \end{cases}$$

Similarly, the variable "lubrication" has three categories, so we need two more indicator variables, such as

$$x_4 = \begin{cases} 1 & \text{lubricant \#1} \\ 0 & \text{otherwise} \end{cases} \qquad x_5 = \begin{cases} 1 & \text{lubricant \#2} \\ 0 & \text{otherwise} \end{cases}$$

(b) This is a model utility test. The hypotheses are H_0: $\beta_1 = \beta_2 = \beta_3 = \beta_4 = \beta_5 = 0$ versus H_a: at least one $\beta_i \neq 0$. From the output, the *F*-statistic is F= 20.67 with a *P-value* of .000. Thus, we strongly reject H_0 and conclude that at least one of the explanatory variables is a significant predictor of springback.

(c) First, find $\hat{y}$ for those settings: $\hat{y}$ = 21.5322 – 0.0033680(1000) – 1.7181(1) – 1.4840(0) – 0.3036(0) + 0.8931(0) = 21.5322 – 0.0033680(1000) – 1.7181 = 16.4461. The error df is 30, so a 95% PI for a new value at these settings is = $16.4461 \pm 2.042\sqrt{(1.18413)^2 + (.524)^2}$ = (13.80, 19.09).

(d) The coefficient of determination in the absence of the lubrication indicators is

$R^2 = 1 - \frac{\text{SSE}}{\text{SST}} = 1 - \frac{48.426}{186.980}$ = .741 or 74.1%. That's a negligible drop in R^2, so we suspect keeping the indicator variables for lubrication regimen is not worthwhile.

More formally, we can test H_0: $\beta_4 = \beta_5 = 0$ versus H_a: $\beta_4 \neq 0$ or $\beta_5 \neq 0$. The F test statistic is F = $\frac{(48.426 - 42.065)/2}{42.065/30}$ = 2.27. The P-value = P($F_{2,30}$ > 2.27) = 0.121 is larger than 0.10, so we fail to reject H_0 at the .10 level. The data does not suggest that lubrication regimen needs to be included so long as BHP and supplier are retained in the model.

(e) R^2 has certainly increased, but that will always happen with more predictors. Let's test the null hypothesis that the interaction terms are <u>not</u> statistically significant contributors to the model. The larger model contributes 4 additional variables: $x_1x_2, x_1x_3, x_1x_4, x_1x_5$. So, the larger model has 30 – 4 = 26 error df, and the F statistic = $\frac{(42.065 - 28.216)/4}{28.216/26}$ = 3.19 . Thus, the P-value =P($F_{.4,26}$ > 3.19) = 0.029. Therefore, we should reject H_0 at the .05 level and conclude that the interaction terms, as a group, do contribute significantly to the regression model.

Section 11.6

45. (a)

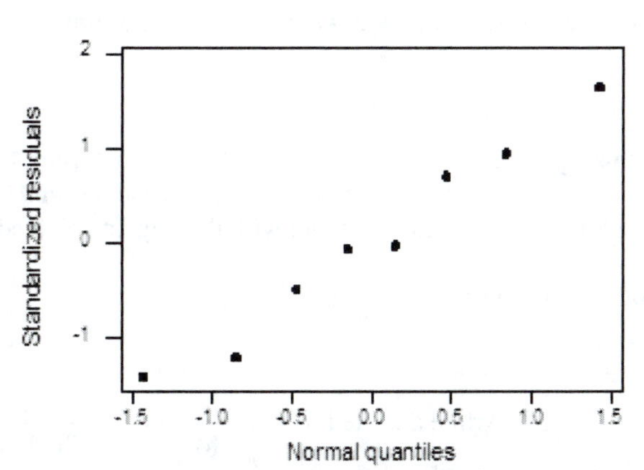

Since the plot of normal quantiles versus standardized residuals looks linear, we would conclude that the standardized residuals are normally distributed.

(b)

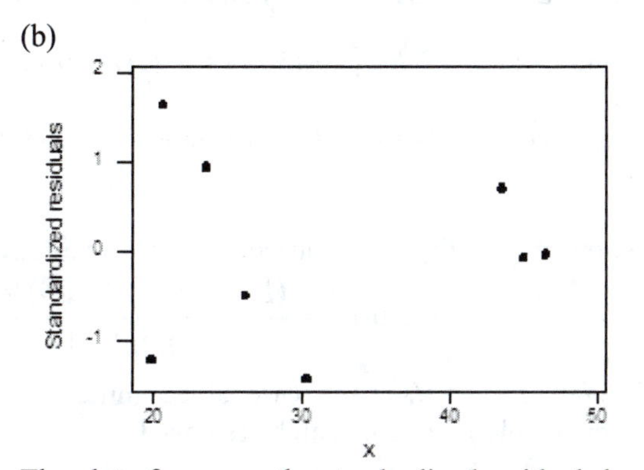

The plot of x versus the standardized residuals has no discernible pattern. So, we would conclude that our simple linear regression model assumptions are being met.

47. (a) Using the equations provided in Section 11.5, the following computations were made.

k	R^2	Adjusted R^2
1	0.6757	0.6462
2	0.9786	0.9744
3	0.9819	0.9758
4	0.9824	0.9736

We would recommend the model with k = 2. As we can see, this model has a substantially higher R^2 adjusted value over the model with k = 1. And, the models with k = 3 and k = 4 give little improvement.

(b) No, a forward selection method would not have considered the k = 2 model described in the example. Forward selection would let x_4 enter the model first and would not delete it at the next stage.

49. The choice of a "best" model seems reasonably clear-cut. The model with 4 variables including all but the summerwood fiber variable would seem best. R^2 is as large as any of the models, including the 5 variable model. R^2 adjusted is at its maximum and CP is at its minimum.

As a second choice, one might consider the model with k = 3 which excludes the summerwood fiber and springwood % variables.

51. The choice of the author is the model using three variables x_3, x_9, x_{10}. It has an adjusted R^2 only slightly smaller than the largest adjusted R^2. The three variable model has two less variables, and hence, two more degrees of freedom for estimation of SSResid than does the model with the largest adjusted R^2.

As a second choice, the two predictor model is also quite good.

53. (a) $R^2 = 1 - \frac{\text{SSE}}{\text{SST}} = 1 - \frac{10.5513}{30.4395} = .653$ or 65.3%, while adjusted $R^2 = 1 - \frac{\text{MSE}}{\text{MST}} = 1 - \frac{10.5513/24}{30.4395/28} = .596$ or 59.6%. Yes, the model appears to be useful.

(b) The null hypothesis is that none of the 10 second-order terms is statistically significant. The F test statistic $= \frac{(10.5513 - 1.0108)/10}{1.0108/14} = 13.21$, and P-value = $P(F_{10,14} > 13.21) < 0.001$. Hence, we strongly reject H_0 and conclude that at least one of the second-order terms is a statistically significant predictor of protein yield.

(c) We want to compare the "full" model with 14 predictors in (b) to a "reduced" model with 5 fewer predictors $(x_1, x_1^2, x_1x_2, x_1x_3, x_1x_4)$. As in (b), we have F test statistic $= \frac{(1.1887 - 1.0108)/4}{1.0108/14} =$ 0.62, and P-value = $P(F_{4,14} > 0.62) = 0.656$. We fail to reject H_0 at any reasonable significance level; therefore, it indeed appears that the five predictors involving x_1 could all be removed.

(d) The "best" models seem to be the 7-, 8-, 9-, and 10-variable models. All of these models have high adjusted R^2 values, low Mallows' CP values, and low s values compared to the other models. The 6-variable model is notably worse than the 7-variable model; the 11-variable model is "on the cusp," in that its properties are slightly worse than the 10-variable model, but only slightly so.

55. H_o: $\beta = 0$ versus H_a: $\beta \neq 0$
The value of the test statistic, z = .73. Its corresponding P-value is .463. Since the P-value is greater than any sensible choice of alpha, we do not reject the null hypothesis. There is insufficient evidence to claim that age has a significant impact on the presence of kyphosis.

57. (a) For x_1 = pillar height to width ratio, H_o: $\beta_1 = 0$ versus H_a: $\beta_1 \neq 0$, $z = 1.878$, P-value = .0604, reject H_o. For x_2 = pillar strength to stress ratio, H_o: $\beta_2 = 0$ versus H_a: $\beta_2 \neq 0$, $z = 2.145$, P-value = .0319, reject H_o. Each of the variables appears to have a significant impact on pillar stability.

(b) The odds of pillar stability changes by the multiplicative factor $e^{2.774} = 16.02$ when x_1 increases by 1 and x_2 remains fixed. The odds of pillar stability changes by the multiplicative factor $e^{5.668} = 289.46$ when x_2 increases by 1 and x_1 remains fixed.

(c) The table of observations with corresponding probabilities and labels is shown below. Based on this, only two observations had a label that did not match actual stability status. The pillar with ID #3 was labeled as "unstable" when in fact it was stable. The pillar with ID #28 was labeled as "stable" when in fact it was unstable.

ID	x1	x2	Stable?	Prob	Label
1	1.8	2.4	Y	0.996	stable
2	1.65	2.54	Y	0.997	stable
3	**2.7**	**0.84**	**Y**	**0.29**	**unstable**
4	3.67	1.68	Y	0.999	stable
5	1.41	2.41	Y	0.988	stable
6	1.76	1.93	Y	0.936	stable
7	2.1	1.77	Y	0.938	stable
8	2.1	1.5	Y	0.765	stable
9	4.57	2.43	Y	1	stable
10	3.59	5.55	Y	1	stable
11	8.33	2.58	Y	1	stable
12	2.86	2	Y	0.998	stable
13	2.58	3.68	Y	1	stable
14	2.9	1.13	Y	0.787	stable
15	3.89	2.49	Y	1	stable

ID	x1	x2	Stable?	Prob	Label
16	0.8	1.37	N	0.041	unstable
17	0.6	1.27	N	0.014	unstable
18	1.3	0.87	N	0.01	unstable
19	0.83	0.97	N	0.005	unstable
20	0.57	0.94	N	0.002	unstable
21	1.44	1	N	0.03	unstable
22	2.08	0.78	N	0.05	unstable
23	1.5	1.03	N	0.041	unstable
24	1.38	0.82	N	0.009	unstable
25	0.94	1.3	N	0.04	unstable
26	1.58	0.83	N	0.017	unstable
27	1.67	1.05	N	0.072	unstable
28	**3**	**1.19**	**N**	**0.872**	**stable**
29	2.21	0.86	N	0.105	unstable

59. (a)

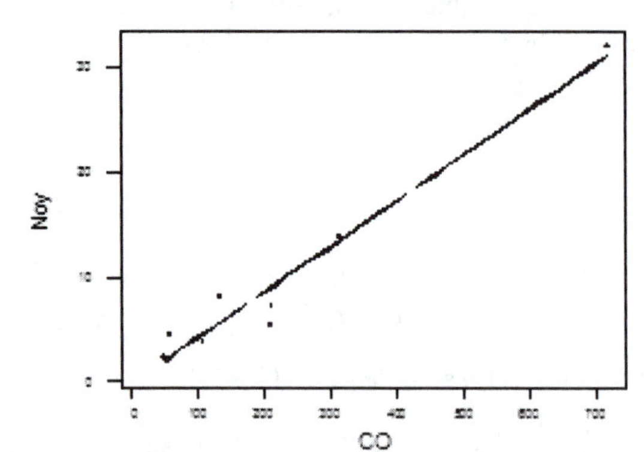

The above analysis was created in Minitab. A simple linear regression model seems to fit the data well. The least squares regression equation is:

$\hat{y} = -.220 +. 0436x$

The model utility test obtained from Minitab produces a t test statistic equal to 12.72. The corresponding P-value is extremely small. So, we have sufficient evidence to claim that ΔCO is a good predictor of ΔNO_y .

(b) $\hat{y} = -.220 + .04362(400) = 17.228$

A 95% prediction interval produced by Minitab is (11.953, 22.503). Since this interval is so wide, it does not appear that ΔNO_y is accurately predicted.

(c) While the large ΔCO value appears to be "near" the least squares regression line, the value has extremely high leverage. The least squares line that is obtained when excluding the value is $\hat{y} = 1.00 + .0346x$. The R^2 value with the value included is 96% and is reduced to 75% when the value is excluded. The value of s_e with the value included is 2.024 and with the value excluded is 1.96.

So, the large ΔCO value does appear to effect our analysis in a substantial way.

61. (a) The five observations made when x = 500 all resulted in different values of y. If the relationship were deterministic, all five of these y values would have been identical.

(b) $\sum x_i = 6075$, $\sum y_i = 371$, $\sum x_i y_i = 226{,}565$, $\sum x_i^2 = 3{,}806{,}125$, $\sum y_i^2 = 13{,}861$ so SS_{xy} = 226,565 - $\frac{1}{10}$(6075)(371) = 1182.5 and SS_{xx} = 3,806,125 - $\frac{1}{10}(6075)^2$ = 115,562.5. Therefore, b = SS_{xy}/SS_{xx} = 1182.5/115,562.5 = .01023256 and a = (371/10) - .01023256(6075/10)= 30.883721. SSTo = 13,861 - $\frac{1}{10}(371)^2$= = 96.90 and SSResid = SSTo - bSS_{xy} = 84.80. The r^2 value is r^2 = 1 - SSResid/SSTo = 1 - 84.80/96.90 = .125. s_e^2 = SSResid/(n-2) = 84.80/(10-2) = 10.6, so s_e = 3.256. The standard error of b is $s_b = s_e/\sqrt{SS_{xy}}$ = 3.256/$\sqrt{115{,}562.5}$ = .009578, so the t-ratio is t = b/s_b = .01023256/.009578 = 1.068, or approximately 1.1. Based on n-2 = 8 d.f., the two-sided P-value associated with t = 1.1 is 2(.152) ≈ .30. Since the P-value exceeds any reasonable value of α, $H_0:\beta = 0$ is <u>not</u> rejected. We can not conclude that the simple linear regression model is useful.

(c) Because the observations on y were made at only two distinct x values, there is no way to distinguish between a simple linear regression model (i.e., a straight line) and any higher-order polynomial model (quadratic, cubic, etc.). For example, to distinguish between a simple linear regression model and a quadratic model requires that observations be made for at least <u>three</u> distinct x values so that the potential curvature in the scatter plot can be observed.

63. (a) The statement is incorrect. r^2 is not the "linear correlation coefficient". r^2 is the coefficient of determination. The linear correlation coefficient is r and $r = \sqrt{.89} = .9434$.

(b)

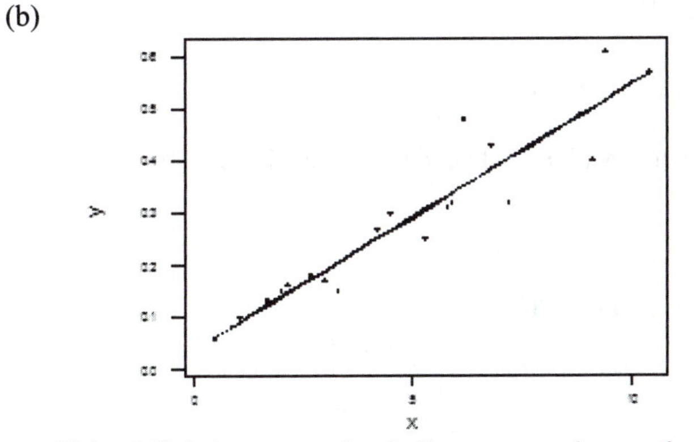

Using Minitab to run a simple linear regression produces the following results.

H_o: $\beta = 0$ versus H_a: $\beta \neq 0$

The value of the t test statistic equals 12.06. The corresponding P-value is extremely small. So, we reject the null hypothesis and conclude that there is a linear relationship between the two variables.

(c)

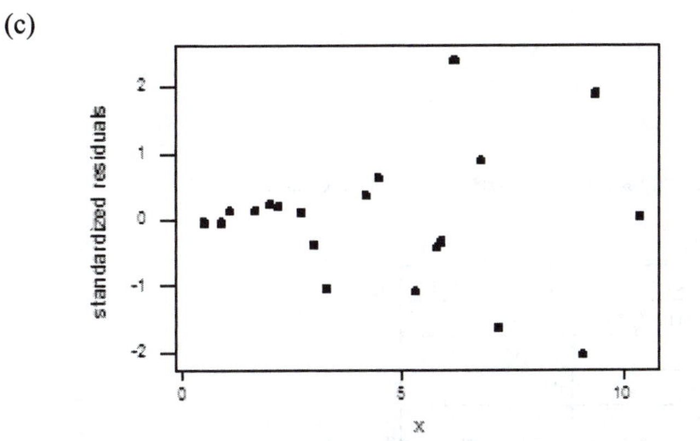

A disturbing pattern is seen in the above plot. As x increases so does the variation in the standardized residuals. This fact is inconsistent with our constant variance assumption of a least squares regression analysis. Remedial action must be taken.

(Note: A transformation may remedy this problem.)

65. The hypotheses are H_0: None of the 6 second order predictors are useful, versus H_a: at least one of the 6 second order terms are useful in predicting oxalic acid production. The second order terms are (x_1x_2, x_1x_3, x_2x_3, and x_1^2, x_3^2, x_3^2). The F test statistic is F = $\frac{(1523351 - 805534)/6}{805534/5}$ = 0.743. Thus, the P-value = $P(F_{6,5} > 3.56) = 0.642 > 0.01$. Hence, we do not reject H_0 and have no evidence to conclude that at least one of the second order terms are useful in predicting oxalic acid production.

67. (a) A point prediction under the specified conditions is:

$$\hat{y}=84.67+.650-.258+.133+.108-.135+.028+.028-.072+.038-.075+.213+.200-.188+.050=85.39$$

(Note: The specified conditions correspond to $x_1 = x_2 = x_3 = x_4 = 1$

The value of the residual for the one observation made under the specified conditions is:

$$(y-\hat{y})=(85.4-85.39)=.01$$

(b) Let z_1, z_2, z_3, z_4 denote the uncoded variables. Then:

$z_1 = .1x_1 + .3$
$z_2 = .1x_2 + .3$
$z_3 = x_3 + 2.5$
$z_4 = 15x_4 + 160$

Algebra produces:

$x_1 = 10z_1 - 3$
$x_2 = 10z_2 - 3$
$x_3 = z_3 - 2.5$
$x_4 = (z_4 - 160)/15$

Substitution yields the following least squares regression coefficients:

Term	Coefficient
Constant	76.437
z_1	-7.35
z_2	9.61
z_3	-.915
z_4	.09632
z_1^2	-13.452
z_2^2	2.798
z_3^2	.02798
z_4^2	-.0003201
z_1z_2	3.750
z_1z_3	-.7500
z_1z_4	.14167
z_2z_3	2.000
z_2z_4	-.1250
z_3z_4	.00333

(c) The full model contains k = 14 variables. The reduced model contains 4 variables.

H_o: $\beta_5 = \ldots = \beta_{14} = 0$ versus

H_o: At least one of the β's is not zero.

Quantities required to compute the test statistic are:

SSResid(full) and SSResid(reduced).

Since SSTo = 17.2567 and R^2(full) = .885, we know SSResid(full) = 1.9845

Since R^2(reduced) = .721, we know SSResid(reduced) = 4.8146

The value of the test statistic is:

$$F=\left[\frac{(4.8146-1.9845)/10}{(1.9845)/(31-15)}\right]=2.28$$

The F critical value at $\alpha = .05$, with num df = 10 and den df = 16 is 2.49.

Since our test statistic is less than the critical value (2.28 < 2.49) we do not reject the null hypothesis. There is not sufficient evidence to claim that the second-order predictors provide useful information beyond what is contained in the four first-order predictors.

69. The following plot of *y* versus *x* suggests that simple linear regression model *may* be appropriate, but a graph of the residuals versus fitted values questions the validity of a simple linear regression model. Fitting higher order models may be more appropriate.

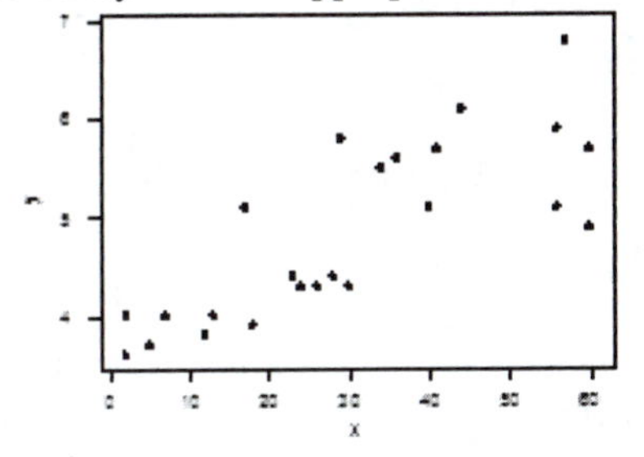

```
The regression equation is
y = 3.70 + 0.0379 x

Predictor        Coef     SE Coef        T       P
Constant       3.6965      0.2159    17.12   0.000
x            0.037895    0.006137     6.17   0.000

S = 0.5525       R-Sq = 63.4%     R-Sq(adj) = 61.7%
```

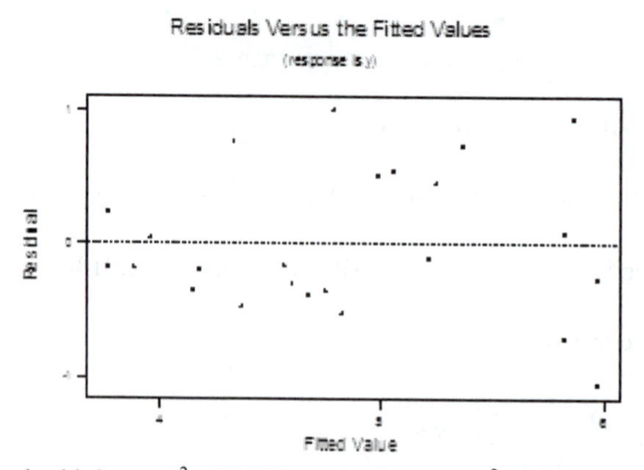

The third order model has the highest R^2 (70.7%) and adjusted R^2 (66.3%). From the second order model, we predict y (at x =30) to be $3.45+.0618(30)-.000377(30^2)=4.9647$. For the third order model, our estimate for y is

$3.94-.045(30)=.0041(30^2)-.000048(30^3)=4.984$. Both models appear to give roughly the same estimate.

```
The regression equation is
y = 3.45 + 0.0618 x -0.000377 x^2

Predictor          Coef      SE Coef          T        P
Constant         3.4465       0.3176      10.85    0.000
x               0.06178      0.02314       2.67    0.014
x^2          -0.0003771    0.0003523      -1.07    0.297

S = 0.5507       R-Sq = 65.3%      R-Sq(adj) = 62.0%
The regression equation is
y = 3.94 - 0.0450 x + 0.00410 x^2 -0.000048 x^3

Predictor          Coef      SE Coef          T        P
Constant         3.9434       0.3961       9.95    0.000
x              -0.04503      0.05992      -0.75    0.461
x^2            0.004099     0.002363       1.74    0.098
x^3         -0.00004840   0.00002529      -1.91    0.070

S = 0.5188       R-Sq = 70.7%      R-Sq(adj) = 66.3%
```

71. (a) The following boxplot shows that the shapes of the ppv for the cracked and uncracked prisms appear to be fairly symmetric. The boxplot further suggests that the ppv for the cracked prisms tend to be greater than the ppv for the uncracked prisms.

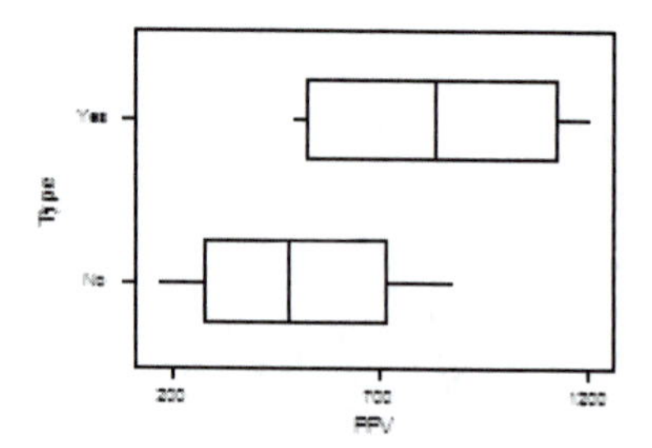

The sample mean and standard deviation for the uncracked prisms are 482.7 and 233.7, respectively, while the sample mean and standard deviation for the cracked prisms are 827.4 and 295.3, respectively. A 95% CI for the difference in these two means is given by $(\bar{x}_1 - \bar{x}_2) \pm t\sqrt{\dfrac{s_1^2}{n_1} + \dfrac{s_2^2}{n_2}}$, where the t value has

$$\text{df} = \frac{(s_1^2/n_1 + s_2^2/n_2)^2}{\dfrac{(s_1^2/n_1)^2}{n_1 - 1} + \dfrac{(s_2^2/n_2)^2}{n_2 - 1}} = \frac{(233.7^2/18 + 295.3^2/12)^2}{\dfrac{(233.7^2/18)^2}{18-1} + \dfrac{(295.3^2/12)^2}{12-1}} = 19.86$$

, so round down to 19 df. So the appropriate t value for a 95% CI with 19 df is 2.093, and so our 95% CI is:

$$(482.7 - 827.4) \pm 2.093\sqrt{\frac{233.7^2}{18} + \frac{295.3^2}{12}} \Rightarrow -344.7 \pm 2.093(101.494)$$

, or (-557.127, -132.273).

(b) Using MINITAB, we can use the best subsets option using the PPV, PPV^2, the indicator variable Crack? (0 if there is no crack present and 1 if there is a crack), and the interaction term PPV*Crack? . The best subsets regression suggests that the single quadratic term PPV^2 is the single most useful predictor. The quadratic regression model, which has the R^2 value of 61.2%, has the equation $\hat{y} = .996719 - .00000001(PPV)^2$. The next most useful single predictor is the PPV term. This simple linear regression model, which has the R^2 value of 57.7%, has the equation $\hat{y} = 1.00161 - .000018(PPV)$. Models involving more than 1 term don't appear to explain the ratio variable any more significantly, since the R^2 values of such models are not much different than the model that simply uses PPV^2 or PPV.

Vars	R-Sq	R-Sq(adj)	C-p	S	PPV	PPV^2	PPV*Crac	Crack?
1	61.2	59.8	-0.8	0.0046880		X		
1	57.7	56.2	1.4	0.0048923	X			
2	61.4	58.6	1.0	0.0047585		X	X	
2	61.4	58.6	1.0	0.0047587		X		X
3	61.5	57.0	3.0	0.0048477	X	X		X
3	61.5	57.0	3.0	0.0048487		X	X	X
4	61.5	55.3	5.0	0.0049435	X	X	X	X